Spectral Models of Random Fields in Monte Carlo Methods

Spectral Models of Random Fields in Monte Carlo Methods

S.M. Prigarin

///VSP///

Utrecht • Boston • Köln • Tokyo - 2001

VSP BV
P.O. Box 346
3700 AH Zeist
The Netherlands

Tel: +31 30 692 5790
Fax: +31 30 693 2081
vsppub@compuserve.com
www.vsppub.com

First published in 2001

ISBN 90-6764-343-2

Printed in The Netherlands by Ridderprint bv, Ridderkerk.

To my beloved Grannies,
Eugenia and Maria

Preface

The proposed spectral models represent a new class of numerical methods aimed at simulation of random processes and fields. The spectral models were developed in the 70-s, and they have appeared to be considerably promising for various applications. In Russia, the papers by Prof. G.A. Mikhailov have given impetus to a large number of investigations on spectral models of random fields. Nowadays the spectral models are extensively used for stochastic simulation in the atmosphere and ocean optics, turbulence theory, analysis of pollution transport for porous media, astrophysics, and other fields of science.

The scalar spectral models and some of their applications are presented in Chapter 1. A new technique of successive refinement of spectral models and conditional spectral models are described here. Chapter 2 deals with vector-valued spectral models. Convergence of spectral models is studied in Chapter 3. Problems of optimization and convergence for functional Monte Carlo estimators are considered in Chapter 4. Moreover, the monograph includes four Appendices, where some auxiliary information is presented and additional problems are discussed. Appendices deal with a) properties and methods of simulation of Gaussian distributions, b) numerical solution of boundary value problems for linear systems of stochastic differential equations, c) interpolation of stationary random sequences and their correlation functions, d) coding of multiplicative pseudo-random number generators in high-level programming languages.

The book is intended for experts in Monte Carlo methods, as well as for senior and post-graduate students, who are interested in the theory and applications of stochastic simulation. For preparing the text, the author used the notes of the lectures he has given at Novosibirsk State University. Knowledge of basic concepts of linear algebra, probability theory and functional analysis is desirable for reading the book. A number of "exercises" and "tasks" are offered for studying on reader's own. Some of the "tasks" contain comparatively difficult and unsolved problems that can be used as subjects of graduate and post-graduate research.

The author would be interested in readers' references that can be sent at the following address or by electronic mail: Institute of Computational Mathematics and Mathematical Geophysics, pr. Lavrentieva, 6, Novosibirsk, 630090, Russia (Email: smp@osmf.sscc.ru).

Acknowledgements. My thanks are due to a number of my teachers, colleagues, and students. I am sincerely grateful to Gennadii Mikhailov for his encouragement and stimulating ideas. I would like to thank my colleagues Tatiana Averina, Klaus Hahn, Stefan Heinrich, Boris Kargin, Klaus Noack, Vasilii Ogorodnikov, Ulrich Oppel, Anton Voitishek, and Gerhard Winkler for close cooperation and fruitful discussions. Special thanks are due to my students at Novosibirsk State University. Finally, I take the opportunity to thank the Russian Foundation for Basic Research and INTAS for the financial support (grants 00-01-00797, 00-05-65456, IR-97-1441).

Akademgorodok, Novosibirsk, Russia Sergei M. Prigarin
September, 2000

Contents

Chapter 1

Approximate modelling of homogeneous Gaussian fields on the basis of spectral decomposition

1.1. Spectral models of random processes and fields

1.1.1. Basic principles of constructing spectral models

Let us consider a real homogeneous random Gaussian field $w(x)$, $x \in \mathbb{R}^k$, with zero mean, unit variance and correlation function $R(x) = \mathbf{M}w(x+y)w(y)$. The spectral representations of the random field and its correlation function are of the form (see, for example, [101])

$$w(x) = \int_{\mathbb{P}} \cos < x, \lambda > \xi(d\lambda) + \int_{\mathbb{P}} \sin < x, \lambda > \eta(d\lambda), \qquad (1.1)$$

$$R(x) = \int_{\mathbb{P}} \cos < x, \lambda > \mu(d\lambda). \qquad (1.2)$$

Here $\xi(d\lambda)$, $\eta(d\lambda)$ are real-valued orthogonal stochastic Gaussian measures on a half-space $\mathbb{P}$ that will be called "spectral space" (i.e., $\mathbb{P}$ is a measurable set such that $\mathbb{P} \cap (-\mathbb{P}) = \{0\}$, $\mathbb{P} \cup (-\mathbb{P}) = \mathbb{R}^k$); $\mu(d\lambda)$ is a spectral measure of the random field $w(x)$ and $< . , . >$ denotes the scalar product in $\mathbb{R}^k$. In this case the following properties are fulfilled:

1) $\mathbf{M}\xi(A) = \mathbf{M}\eta(A) = 0$;
2) $\mathbf{M}\xi(A)\eta(A) = 0$;
3) $\mathbf{M}\xi^2(A) = \mathbf{M}\eta^2(A) = \mu(A)$;
4) if $A \cap B = \emptyset$, then
$$\mathbf{M}\xi(A)\xi(B) = \mathbf{M}\eta(A)\eta(B) = 0,$$

where A and B stand for measurable subsets of $\mathbb{P}$.

The main idea underlying the methods of constructing spectral models is to take an approximation of stochastic integral (1.1) as a numerical model of the random field $w(x)$.

A simple spectral model may be constructed as follows. Let us fix some splitting of the "spectral space" $\mathbb{P}$:

$$\mathbb{P} = \sum_{j=1}^{n} Q_j, \quad Q_j \cap Q_i = \emptyset \quad \text{for } i \neq j.$$

As an approximation for (1.1) we consider

$$w_n(x) = \sum_{j=1}^{n} \mu^{1/2}(Q_j)\,[\xi_j \cos < \lambda_j, x > + \eta_j \sin < \lambda_j, x >]\,. \tag{1.3}$$

Here ξ_j, η_j are independent Gaussian random variables,

$$\mathbf{M}\xi_j = \mathbf{M}\eta_j = \mathbf{M}\xi_j\eta_k = 0, \quad \mathbf{M}\xi_j^2 = \mathbf{M}\eta_j^2 = 1,$$

while vectors $\lambda_j \in \mathbb{P}$ belong to the corresponding sets Q_j. The random fields $w_n(x)$ in formula (1.3) are homogeneous Gaussian fields with the correlation function

$$R_n(x) = \sum_{j=1}^{n} \mu(Q_j) \cos < \lambda_j, x > .$$

Remark. Formula (1.3) may be reduced to an equivalent, but more effective for modelling, form:

$$w_n(x) = \sum_{j=1}^{n} \mu^{1/2}(Q_j)\; r_j \;\cos(< \lambda_j, x > + \varphi_j), \tag{1.4}$$

$$r_j = \left(\xi_j^2 + \eta_j^2\right)^{1/2} = (-2 \ln \alpha_j)^{1/2},$$

$$\varphi_j = \arg(\xi_j + i\eta_j) = 2\pi\alpha_j',$$

where r_j, φ_j are independent random variables: φ_j are uniformly distributed in the interval $(0, 2\pi)$, r_j are random variables with Rayleigh distribution, α_j and α_j' are independent random variables uniformly distributed in the interval (0,1).

Another important concept used in constructing spectral models is the *spectrum randomization* principle. To illustrate the principle, let us consider a harmonic

$$r\cos(<\lambda, x> +\varphi) \tag{1.5}$$

where the amplitude r, frequency λ, and phase φ are random and mutually independent. If $\mathbf{M}r^2 = 1$, the value of φ is uniformly distributed in the interval $(0, 2\pi)$, and the frequency λ is distributed according to measure μ, then one can easily show that the field (1.5) is homogeneous with the spectral measure μ. Summing the independent realizations of random harmonics (1.5) and normalizing the sum, we obtain the asymptotically Gaussian sequence of homogeneous random fields with the spectral measure μ:

$$w_n(x) = (1/n)^{1/2} \sum_{j=1}^{n} r_j \cos(<\lambda_j, x> +\varphi_j). \tag{1.6}$$

Such models were discussed in [110].

Thus, the spectrum randomization principle (the frequencies λ_j are randomly chosen, according to the spectral measure μ) allows one to reproduce exactly the correlation function of the field (1.1), which is of particular importance for solving problems by Monte Carlo method, when independent realizations of a random field are simulated many times in order to compute some functionals.

The most promising spectral model for Monte Carlo methods was proposed in [71]. The model combines the above principles: spectral space partitioning and spectrum randomization. In this case the modelling formula coincides with (1.3), the essential distinction being that the vectors λ_j are chosen randomly in the corresponding domains Q_j according to the distributions generated by the spectral measure μ. As in the case of model (1.6), the spectrum randomization principle allows one to reproduce exactly the correlation function of the limit field (1.1). A possibility to vary the way of partitioning of the spectral space makes the model more flexible than model (1.6) and enables one to obtain a higher adequacy, using a lesser number of harmonics.

We call models (1.3) and (1.6), constructed with the spectrum randomization principle, the *randomized spectral models.*

Remark. Random variables ξ_j, η_j in formula (1.3) can be non-Gaussian as well. Non-randomized model is non-Gaussian in this case and one-dimensional distribution of randomized spectral model is non-Gaussian, too. But the property of the models to be asymptotically Gaussian is retained. In particular, one can use spectral models with random variables ξ_j, η_j which take values +1, –1 with probability 0.5 in formula (1.3) and, correspondingly, $r_j = 2^{1/2}$ in formulas (1.4), (1.6). However, in what follows we shall assume that random variables ξ_j, η_j are Gaussian if the reverse is not mentioned.

1.1.2. Generalized scheme. About numerical analysis of the error

Let us consider a generalized randomized scheme of approximate modelling of a homogeneous Gaussian random field $w(x)$ with zero mean, unit variance and spectral measure μ:

$$w_n(x) = \sum_{j=1}^{n} a_j \left[\xi_j \cos < \lambda_j, x > + \eta_j \sin < \lambda_j, x >\right], \qquad (1.7)$$

where ξ_j, η_j are independent random variables,

$$\mathbf{M}\xi_j = \mathbf{M}\eta_j = \mathbf{M}\xi_j\eta_k = 0, \quad \mathbf{M}\xi_j^2 = \mathbf{M}\eta_j^2 = 1,$$

λ_j are random vectors independent of $(\xi_j, \eta_j)_{j=1\ldots n}$ and distributed according to such probability measures μ_j that

$$\mu(d\lambda) = \sum_{j=1}^{n} a_j^2 \mu_j(d\lambda), \quad \sum_{j=1}^{n} a_j^2 = 1. \qquad (1.8)$$

Condition (1.8) guarantees the coincidence of spectra for the field $w(x)$ and the approximate model (1.7), while from convergence $\max_{j\leq n} a_j \underset{n\to\infty}{\longrightarrow} 0$ it follows that the approximate model is asymptotically Gaussian (cf., for example, [31]. Note that random vectors λ_j, distributed according to measures μ_j, can depend on each other.

Efficiency of numerical algorithms which use simulation of random fields can highly depend on the choice of an approximate model of a random field. General problems of convergence of spectral models are discussed in Chapter 3. Below we consider the question: how to measure a spectral model error?

First we deal with non-randomized model (1.3). As an error $\Delta(w_n,w)$ we can take some "distance" between spectral measures of the random field w and its approximate model w_n or "distance" between the corresponding correlation functions. A distinctive feature of a non-randomized model is that its spectral measure is discrete, while correlation function is almost-periodic.

In the next item we consider some examples of spectral models for Gaussian stationary processes and the following quantity is used as an error

$$\Delta(w_n,w) = \|q(x)[R(x) - R_n(x)]\|\,.$$

Here

$$q(x) = \begin{cases} 1 - |x|/T, & |x| < T, \\ 0, & |x| \geq T, \end{cases}$$

and the norm is taken for space C or L^p. In particular, for L^p, $p = 2$, we have

$$\Delta(w_n,w) = \int_{-\infty}^{+\infty} q^2(x)|R(x) - R_n(x)|^2 dx = 2\pi \int_{-\infty}^{+\infty} |f(\lambda) - f_n(\lambda)|^2 d\lambda,$$

$$q(x)R_{(n)}(x) = \int_{-\infty}^{+\infty} \exp(i\lambda x) f_{(n)}(\lambda) d\lambda,$$

$$f_{(n)}(\lambda) = (2\pi)^{-1} \int_{-\infty}^{+\infty} \exp(-i\lambda x) q(x) R_{(n)}(x) dx,$$

where the distribution density $f_{(n)}(\lambda)$ is is the convolution of probability density

$$\frac{T}{2\pi}\left[\frac{\sin(T\lambda/2)}{T\lambda/2}\right]^2 = \frac{1}{\pi}\,\frac{1-\cos T\lambda}{T\lambda^2}, \qquad \lambda \in (-\infty, +\infty),$$

with the spectral measure $\mu_{(n)}$.

Concerning realizations of randomized spectral models, we can imagine that they are homogeneous random fields with discrete spectra concentrated at points λ_j, which are chosen randomly. In this case as an error of generalized spectral model (1.7) we consider the following quantities

$$[\mathbf{M}\Delta^r(w_n, w)]^{1/r},$$

where $\Delta(w_n, w) = \|q(x)[R(x) - R_n(x)]\|$ is a random variable, $R_n(x) = \sum a_j^2 \cos < \lambda_j, x >$ and mathematical expectation is taken with respect to joint distribution of vectors λ_j.

Remark. Assume that a sequence of spectral models w_n and the limit field w belong to a a functional space with a metric ρ. Let us suppose in addition that spectral models w_n are constructed on the same probability space (see section 1.2 below). Then the quantity

$$\mathbf{M}\rho(w_{n_\infty}, w_n),$$

can be used to estimate numerically an error of spectral model w_n. This approach is reasonable if $\rho(w, w_n) \to 0$ for $n \to \infty$ and n_∞ is large enough for approximate equality

$$\rho(w, w_n) \approx \rho(w_{n_\infty}, w_n).$$

But, evidently, this method can be time-consuming and it is not appropriate for all spectral models.

1.1.3. Examples of spectral models of stationary processes

In this item we use the following notation: $w(x)$ is a simulated stationary process with zero mean and unit variance, $x \in \mathbb{R}$; $R(x)$ is a correlation function of the process $w(x)$ and $\mu(d\lambda)$ is its spectral measure concentrated in $[0, +\infty)$; $w_n(x)$ is an approximate spectral model of the random process $w(x)$,

$$w_n(x) = \sum_{j=1}^{n} a_j[\xi_j \cos(\lambda_j x) + \eta_j \sin(\lambda_j x)],$$

where ξ_j, η_j are independent standard normal random variables; $R_n(x)$ and $\mu_n(d\lambda)$ are correlation function and spectral measure of the process $w_n(x)$. It is fulfilled $R_n(x) = \sum_{j=1}^{n} a_j^2 \cos(\lambda_j x)$ for non-randomized model and $R_n(x) = R(x)$ for randomized one. By $F(\lambda)$ and $f(\lambda)$ we denote spectral function and spectral density of the process $w(x)$:

$$dF(\lambda) = \mu(d\lambda) = f(\lambda)d\lambda,$$

$$R(x) = \int_0^{+\infty} \cos(\lambda x) f(\lambda) d\lambda, \quad f(\lambda) = \frac{2}{\pi} \int_0^{+\infty} \cos(\lambda x) R(x) dx.$$

Example 1.

$$f(\lambda) = \begin{cases} 1/A, & \lambda \in [0, A], \\ 0 & \text{otherwise,} \end{cases} \qquad R(x) = \frac{\sin(Ax)}{Ax}.$$

Let us set $a_j^2 = 1/n$ and consider the following spectral models.

Model D (non-randomized): $\lambda_j = (j - 0.5)A/n$

Model $R1$ (randomized): λ_j are mutually independent and uniformly distributed in corresponding intervals $[(j-1)A/n, jA/n]$.

Model $R2$ (randomized): λ_j are mutually independent and uniformly distributed in the interval [0,A].

To measure the error of the spectral models we use

$$\delta = \mathbf{M} \left\| q(x) \left[R(x) - \sum_{j=1}^{n} a_j^2 \cos(\lambda_j x) \right] \right\|,$$

where

$$q(x) = \begin{cases} 1 - |x|/T, & |x| < T, \\ 0, & |x| \geq T, \end{cases}$$

and mathematical expectation is taken with respect to a joint distribution of values λ_j, $j = 1, ..., n$. By $\delta_1, \delta_2, \delta_c$ we denote the values of errors for norms in spaces $L_1[0,T]$, $L_2[0,T]$, $C[0,T]$ respectively. The values of errors of spectral models for $T = 1$, $A = 10$, $n = 10$ are presented in Table 1.1. Out of the randomized models the errors are least for Model $R1$.

Table 1.1

Model	δ_1	δ_2	δ_c
D	0.0005	0.0006	0.0010
$R1$	0.0085	0.0108	0.0202
$R2$	0.0618	0.0800	0.1495

Remark. An error of a randomized model can be significantly larger in comparison with that of a non-randomized model. It may be considered as the price for a helpful property of randomized models – exact reproducing of spectrum and correlation function of simulated field according to "ensemble of realizations", which is often essential for solving applied problems by Monte Carlo method. Therefore, the considered quantities can be recommended for measuring errors either in class of randomized models or in class of non-randomized models but not for comparing models belonging to different classes.

Example 2.

$$f(\lambda) = \begin{cases} 2(1-\lambda/A)/A, & \lambda \in [0, A], \\ 0 & \text{otherwise}, \end{cases}$$

$$R(x) = 2\frac{1-\cos Ax}{(Ax)^2} = \left[\frac{\sin(Ax/2)}{Ax/2}\right]^2,$$

$$F(\lambda) = -\lambda^2/A^2 + 2\lambda/A, \quad \lambda \in [0, A],$$

$$F^{-1}(c) = A\left(1 - \sqrt{1-c}\right), \quad c \in (0, 1).$$

Let us consider the following spectral models.

Model $D1$ (non-randomized):

$$a_j^2 = F(jA/n) - F((j-1)A/n), \quad \lambda_j = (j-0.5)A/n.$$

Model $D2$ (non-randomized): $a_j^2 = 1/n$,

$$\lambda_j = \frac{1}{2}\left[F^{-1}((j-1)/n) + F^{-1}(j/n)\right].$$

Model $R1$ (randomized):

$$a_j^2 = F(jA/n) - F((j-1)A/n),$$

λ_j are mutually independent and distributed in the corresponding intervals $[(j-1)A/n, jA/n]$ according to the density $f(\lambda)$.

Model $R2$ (randomized): $a_j^2 = 1/n$, λ_j are independent and distributed in the corresponding intervals $(F^{-1}((j-1)/n),\ F^{-1}(j/n))$ according to the density that is proportional to $f(\lambda)$.

Model $R3$ (randomized): $a_j^2 = 1/n$, λ_j are independent and identically distributed according to the density $f(\lambda)$ in the interval $(0, A)$.

Remark. Random variable λ distributed according to a density $f(\lambda)$ with the constraint $\lambda \in [c, d]$ can be simulated by formula

$$\lambda = F^{-1}\left(F(c) + \alpha\left(F(d) - F(c)\right)\right),$$

where α is a random variable uniformly distributed in the interval [0,1].

The estimates of errors for different spectral models are presented in Table 1.2 for $T = 1$, $A = 60$, $n = 10$. In this case the models without partitioning of spectrum and with partitioning into segments with equal weights are the most promising among randomized models.

Table 1.2

Model	δ_1	δ_2	δ_c
$D1$	0.001	0.002	0.005
$D2$	0.028	0.037	0.086
$R1$	0.052	0.071	0.176
$R2$	0.040	0.055	0.149
$R3$	0.041	0.055	0.149

1.1.4. Examples of spectral models for isotropic fields on a plane

Correlation function of any isotropic field $w(x, y)$ on a plane is represented in the form

$$\begin{aligned} B(r) &= \mathbf{M}[w(0,0)w(x,y)] - \mathbf{M}w(0,0)\mathbf{M}w(x,y) = \\ &= \sigma^2 \int_0^{+\infty} J_0(\rho r)\nu(d\rho), \end{aligned} \tag{1.9}$$

where $r^2 = x^2 + y^2$, σ^2 is a variance of the field, J_0 is the Bessel function of the first kind, $\nu(d\rho)$ is a "radial" spectral measure on $[0, \infty)$. In what follows we assume that measure ν has a density:

$$\nu(d\rho) = g(\rho)d\rho, \quad g(\rho) = \rho \int_0^{+\infty} x J_0(\rho x) B(x) dx.$$

By G we denote distribution function of measure ν:

$$dG(\rho) = \nu(d\rho) = g(\rho)d\rho, \quad G(\rho) = \rho \int_0^{+\infty} J_1(\rho x)B(x)dx.$$

A list of spectral densities and corresponding correlation functions of isotropic fields on a plane is presented in Table 1.3.

Table 1.3. Correlation functions $B(t)$, radial spectral densities $g(z)$ for isotropic fields on the plane and correspondent spectral densities $f(x)$ for random processes

$f(x)/2, \quad x > 0$	$B(t), \quad t > 0$	$g(z), \quad z > 0$
$\frac{1}{\pi\sqrt{a^2 - x^2}}, \quad x < a$	$J_0(at)$	$\delta(a - z)$
$\frac{1}{2a}, \quad x < a$	$\frac{\sin(at)}{at}$	$\frac{z}{a\sqrt{a^2 - z^2}}, \quad z < a$
$\frac{2}{\pi a^2}\sqrt{a^2 - x^2}, \quad x < a$	$2\frac{J_1(at)}{at}$	$\frac{2}{a^2}z, \ z < a$
$\frac{3}{4a^3}(a^2 - x^2), \quad x < a$	$3\frac{\sin(at) - at\cos(at)}{(at)^3}$	$\frac{3}{a^3}z\sqrt{a^2 - x^2}, \ z < a$
$C(a^2 - x^2)^{\nu - 1/2}, x < a$ $C = \frac{\Gamma(\nu + 1)}{\sqrt{\pi}a^{2\nu}\Gamma(\nu + 1/2)}$ $\nu > 0,$	$2^\nu\Gamma(\nu + 1)\frac{J_\nu(at)}{(at)^\nu}$	$\frac{2\nu}{a^{2\nu}}z(a^2 - z^2)^{\nu - 1},$ $z < a$
$C(a^2 + x^2)^{-\nu - 1/2}, \nu > 0$ $C = \frac{a^{2\nu}\Gamma(\nu + 1/2)}{\sqrt{\pi}\Gamma(\nu)}$	$\frac{2^{-\nu+1}}{\Gamma(\nu)}(at)^\nu K_\nu(at)$	$2\nu a^{2\nu}z(z^2 + a^2)^{-\nu - 1}$
$\frac{a}{\pi(a^2 + x^2)}$	$\exp(-at)$	$az(z^2 + a^2)^{-3/2}$
$\frac{2a^3}{\pi(a^2 + x^2)^2}$	$\exp(-at)(1 + at)$	$3a^3z(z^2 + a^2)^{-5/2}$

Continuation of Table 1.3

$\frac{a}{\pi}K_0(ax)$	$\frac{a}{\sqrt{a^2+t^2}}$	$a\exp(-az)$
$\frac{a}{2}\exp(-ax)$	$\frac{a^2}{a^2+t^2}$	$a^2 zK_0(az)$
$\frac{a}{\pi}(ax)K_1(ax)$	$\frac{a^3}{(a^2+t^2)^{3/2}}$	$a(az)\exp(-az)$
$\frac{a}{4}\exp(-ax)(1+ax)$	$\frac{a^4}{(a^2+t^2)^2}$	$\frac{a}{2}(az)^2K_1(az)$
$\frac{a}{3\pi}(ax)^2K_2(ax)$	$\frac{a^5}{(a^2+t^2)^{5/2}}$	$\frac{a}{3}(az)e^{-az}(1+az)$
$C(ax)^\nu K_\nu(ax), \nu>-\frac{1}{4}$ $C=\frac{a}{2^\nu\sqrt{\pi}\Gamma(\nu+1/2)}$	$\frac{a^{2\nu+1}}{(a^2+t^2)^{\nu+1/2}}$	$\frac{a\sqrt{2}(az)^{\nu+1/2}}{\Gamma(\nu+1/2)2^\nu}K_{\nu-1/2}(az)$
$\frac{1}{a^2}(a-x), \quad x<a$	$\frac{2}{a^2t^2}(1-\cos(at))$	$\frac{2z}{a^2}\operatorname{arch}(a/z), \quad z<a$
$\sqrt{a/\pi}\exp(-ax^2)$	$\exp\left(-\frac{t^2}{4a}\right)$	$2az\exp(-az^2)$
$\frac{\sqrt{a}}{\pi^{3/2}}\exp\left(-\frac{ax^2}{4}\right)\times$ $\times K_0\left(\frac{ax^2}{2}\right)$	$\exp\left(-\frac{t^2}{8a}\right)I_0\left(\frac{t^2}{8a}\right)$	$2\sqrt{a/\pi}\exp(-az^2)$

Note. Here we assume that

$$f(x)=\frac{2}{\pi}\int_0^\infty B(u)\cos(ux)du, \quad f(-x)=f(x),$$

$$\int_0^\infty f(x)dx=\int_{-\infty}^\infty f(x)/2\,dx=\int_0^\infty g(z)dz=1=B(0).$$

The spectral density on the plane corresponding to the radial spectral density $g(z)$ has the form $p(x,y)=(2\pi)^{-1}(x^2+y^2)^{-1/2}g\left((x^2+y^2)^{1/2}\right)$.

Assume that $0 = R_0 < R_1 < \ldots < R_{N-1} < R_N = \infty$. Let us consider the following version of spectral model for isotropic field $w(x, y)$ with mean zero:

$$w^*(x,y)=\sigma \sum_{n=1}^{N} c_n M_n^{-1/2} \sum_{m=1}^{M_n} (-2\ln \alpha_{nm})^{1/2} \times$$
$$\times \cos\left[(x\rho_{nm}\cos\omega_{nm} + y\rho_{nm}\sin\omega_{nm}) + 2\pi\beta_{nm}\right], \qquad (1.10)$$

where

$$c_n^2 = \int_{R_{n-1}}^{R_n} g(\rho)d\rho; \quad \omega_{nm} = \pi(m - \gamma_{nm})/M_n;$$

ρ_{nm} are random variables distributed in $[R_{n-1}, R_n)$ with density $g(\rho)/c_n^2$; α_{nm}, β_{nm}, γ_{nm} are independent random variables uniformly distributed in $[0, 1]$.

The algorithm of simulation involves the creation of the following arrays

$$A(n,m) = \sigma c_n(-2\ln \alpha_{nm}/M_n)^{1/2}, \quad D(n,m) = 2\pi\beta_{nm},$$
$$B(n,m) = \rho_{nm}\cos\omega_{nm}, \quad C(n,m) = \rho_{nm}\sin\omega_{nm},$$

and subsequent calculation of the value of the field at the required point (x, y) by the formula

$$w^*(x,y) = \sum_{n=1}^{N}\sum_{m=1}^{M_n} A(n,m)\cos\left[B(n,m)x + C(n,m)y + D(n,m)\right].$$

Model (1.10) corresponds to the splitting of the spectral space into rings and the partitioning of each ring into the segments of equal magnitude. The correlation function of model (1.10) is equal to (1.9), and if $\sum_{n=1}^{N} M_n \to \infty$ and $\max_{n\le N}(c_n^2/M_n) \to 0$, then the field $w^*(x, y)$ is asymptotically Gaussian.

Some other versions of the spectral model are possible. In particular, the following changes in (1.10) (both separately and in combination) are acceptable

$$1)\ \rho_{n1} = \rho_{n2} = \ldots = \rho_{nM_n} = \rho_n, \qquad 2)\ \gamma_{nm} = \gamma_n, \quad 3)\ \gamma_{nm} = \gamma, \tag{1.11}$$

where ρ_n, γ_n, γ are independent random variables with corresponding distributions. These replacements imply dependence of two-dimensional random vectors $\lambda_j = (\rho_{nm} \cos \omega_{nm}, \rho_{nm} \sin \omega_{nm})$ in representation (1.7).

Algorithmically simpler, but less flexible is model (1.10), where $c_n = N^{-1/2}$ and $\rho_{nm} = \rho_n$ are independent and distributed on the whole semi-axis $[0, \infty)$ with density $g(\rho)$. In this case the splitting of spectral space is performed only partially, in the directions of vectors λ_j.

The choice of a version of the simulation algorithm and its parameters is specified by the aim to represent more or less in detail the corresponding parts of spectrum. The choice affects the character of realizations $w^*(x, y)$.

Example 3. (Figure 1.1).

For approximate modelling of a Gaussian homogeneous isotropic random field $w(x, y)$ with zero mean, unit variance and correlation function $B(r) = J_0(\rho r)$ we consider the following spectral model:

$$w_M^{(\rho)}(x, y) = M^{-1/2} \sum_{m=1}^{M} \Big[\xi_m \cos(x\rho \cos \omega_m + y\rho \sin \omega_m) + \\ + \eta_m \sin(x\rho \cos \omega_m + y\rho \sin \omega_m)\Big], \tag{1.12}$$

where ξ_m, η_m are independent standard normal variables, $\omega_m = \pi(m - \alpha')/M$, α' is a random variable uniformly distributed in the interval $(0, 1)$. Model (1.12) can be written in the equivalent form, which is more effective for computation:

$$w_M^{(\rho)}(x, y) = M^{-1/2} \sum_{m=1}^{M} (-2 \ln \alpha_m)^{1/2} \times \\ \times \cos(x\rho \cos \omega_m + y\rho \sin \omega_m + 2\pi\beta_m),$$

where α_m, β_m are independent and uniformly distributed in $[0, 1]$.

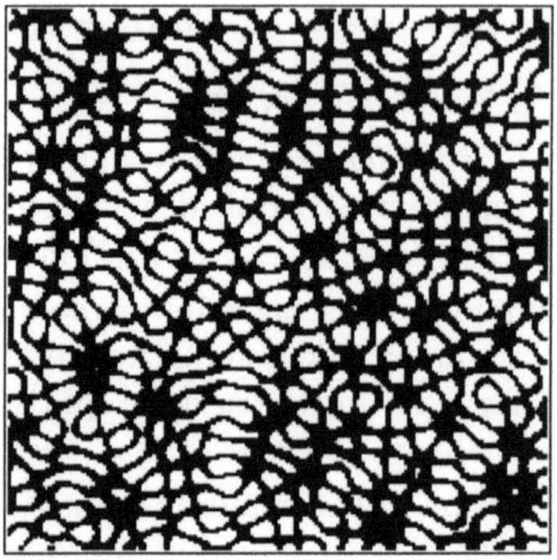

Figure 1.1. Realization of spectral model of a homogeneous isotropic random field with correlation function $J_0(cr)$

In addition to this model, which will be called Model $R1$, we consider two other randomized models with different ways for simulation of ω_m.

Model $R2$:

$$\omega_m = \pi(m - \alpha'_m)/M.$$

Model $R3$ (without splitting the spectrum):

$$\omega_m = \pi\alpha'_m.$$

Random variables α'_m in models $R2$ and $R3$ are mutually independent and uniformly distributed in the interval $(0, 1)$.

We denote the values of errors

$$\mathbf{M}\left\| R(x,y) - M^{-1/2}\sum_{m=1}^{M}\cos(x\rho\cos\omega_m + y\rho\sin\omega_m)\right\|,$$

by δ_1, δ_2, δ_c for norms in spaces $L_1[0,T]$, $L_2[0,T]$ and $C[0,T]$ respectively. Here $R(x,y){=}J_0(\rho\sqrt{x^2{+}y^2})$, while mathematical expectation is taken with respect to joint distribution of random variables ω_m; $m{=}1,\ldots,M$. Table 1.4 presents the errors of spectral models $R1 - R3$ for $T = 1$, $\rho = 10$, $M = 10$.

Randomized model $R1$ has much smaller errors because of splitting spectrum and, particularly, because of dependence between random vectors $\lambda_m = (\rho\cos\omega_m, \rho\sin\omega_m)$.

Table 1.4

Model	δ_1	δ_2	δ_c
$R1$	0.00006	0.00028	0.0031
$R2$	0.081	0.104	0.285
$R3$	0.169	0.207	0.520

An arbitrary isotropic field on a plane can be represented as a superposition of isotropic fields with correlation functions $J_0(\rho\sqrt{x^2+y^2})$. Therefore, an approximate model of an arbitrary isotropic field on a plane can be built as a weighted sum $\sum_\rho a_\rho w_M^{(\rho)}(x,y)$ of spectral models of the form (1.12) (cf. (1.10), (1.11)).

Example 4. (Figure 1.2).

$$B(r) = \exp(-ar), \quad g(\rho) = a\rho(\rho^2+a^2)^{-3/2}, \quad \rho > 0,$$
$$G(\rho) = 1 - a(a^2+\rho^2)^{-1/2}, \quad \rho \geq 0,$$
$$G^{-1}(c) = a\left(\frac{1}{(1-c)^2} - 1\right)^{1/2}, \quad c \in [0,1).$$

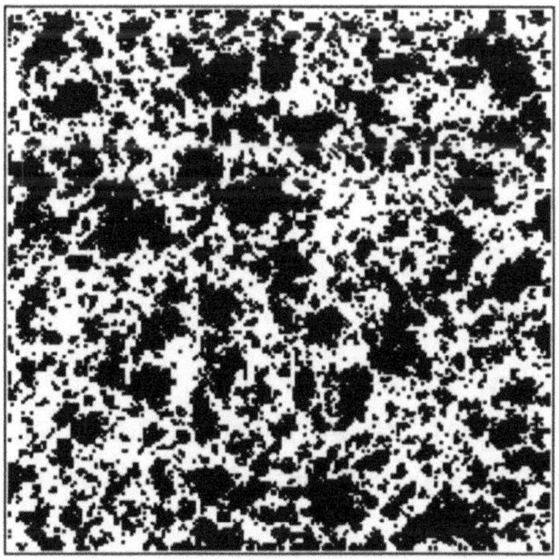

Figure 1.2. Realization of spectral model of a homogeneous isotropic random field with exponential correlation function

Example 5. (Figure 1.3)

$$B(r) = \frac{\sin(ar)}{ar}, \quad g(\rho) = \begin{cases} \frac{\rho}{a(a^2-\rho^2)^{1/2}}, & \rho \in [0,a], \\ 0 & \text{otherwise,} \end{cases}$$

$$G(\rho) = 1 - (a^2-\rho^2)^{1/2}/a, \quad \rho \in [0,a],$$

$$G^{-1}(c) = a\left(1-(1-c)^2\right)^{1/2}, \quad c \in [0,1).$$

Below we present relations between spectral measure

$$F(d\lambda) = f(\lambda)d\lambda, \quad f(\lambda) = \frac{2}{\pi}\int_0^\infty B(r)\cos(\lambda r)dr, \quad \lambda \geq 0,$$

of a stationary random process, which is obtained as a trace on a straight line of a homogeneous isotropic field on $\mathbb{R}^2$, and a radial spectral measure $G(d\rho) = g(\rho)d\rho$ of the isotropic random field (for details see [83] and references presented there):

$$f(\lambda) = \frac{2}{\pi}\int_\lambda^\infty (\rho^2-\lambda^2)^{-1/2}dG(\rho), \quad \lambda \geq 0,$$

$$F(\lambda) = \frac{2}{\pi}\int_0^{\pi/2} G(\lambda/\sin\theta)d\theta = G(\lambda) + \frac{2}{\pi}\int_\lambda^\infty \arcsin(\lambda/\rho)dG(\rho), \ \lambda \geq 0,$$

$$1 - G(\rho) = \rho\int_0^{\pi/2} f(\rho/\sin\theta)\sin^{-2}(\theta)d\theta,$$

$$g(\rho) = \rho\left[(R^2-\rho^2)^{-1/2}f(R) - \int_\rho^R f'(\lambda)(\lambda^2-z^2)^{-1/2}d\lambda\right], \ \rho > 0.$$

In the last equation we assume that $f(\lambda) > 0$ for $\lambda \in [0,R)$ and $f(\lambda) = 0$ for $\lambda \in (R,+\infty)$. In particular, these formulas imply that if the radial spectrum has a power density, i. e. the following is fulfilled

$$g(\rho) = \begin{cases} \text{const}\,\rho^{-r}, & \rho > \rho^*, \\ 0 & \rho \in [0,\rho^*), \end{cases} \quad (\rho^* > 0, \quad r > 1),$$

then the spectral density $f(\lambda)$ is a power function, too, with the same index for $\lambda \geq \rho^*$:

Figure 1.3. Realization of spectral model of a homogeneous isotropic random field with correlation function $\sin(cr)/(cr)$

$$G(\rho) = c_1 \left[1 - (\rho/\rho^*)^{-r+1}\right], \quad \rho > \rho^*,$$

$$F(\lambda)=\frac{2}{\pi}\int_0^{\pi/2} c_1 \left[1-\left(\frac{\lambda}{\rho^* \sin\theta}\right)^{-r+1}\right] d\theta=c_2+c_3\lambda^{-r+1}, \quad \lambda > \rho^*,$$

where c_1, c_2, c_3 are constants.

1.1.5. Spectral models for isotropic fields in three-dimensional space

Let $w(x)$, $x \in \mathbb{R}^3$, be a homogeneous isotropic random field with zero mean and unit variance. Correlation function of the field $w(x)$ can be represented as

$$\mathbf{M}w(x)w(0) = B(r) = \int_0^\infty \frac{\sin(\gamma r)}{\gamma r} dH(\gamma), \tag{1.13}$$

where $r = \|x\|$ is Euclidean norm in $\mathbb{R}^3$, $H(\gamma)$ is the function of probability distribution in $[0, +\infty)$,

$$H(\gamma) = \frac{2}{\pi}\int_0^\infty [\sin(\gamma r) - (\gamma r)\cos(\gamma r)] \frac{B(r)}{r} dr.$$

The correlation function $B(r)$ of a homogeneous isotropic field on $\mathbb{R}^3$ has the following properties (cf., for example, [124], [64]):

1) $B(r) \geq -1/3$;

2) function $B(r)$ is differentiable in $(-\infty, +\infty)$.

Let us assume that distribution $dH(\gamma)$ has a density $h(\gamma)$, $h(\gamma)d\gamma = dH(\gamma)$. Inverse transformation to (1.13) for the density is the following

$$h(\gamma) = \frac{2}{\pi} \int_0^\infty (\gamma r) \sin(\gamma r) B(r) dr.$$

Functions $h(\gamma)$ and $H(\gamma)$ will be called *radial spectral density* and *radial spectral function* of a homogeneous isotropic random field in a three-dimensional space, while corresponding measure will be referred to as *radial spectral measure.*

Example 6. Let $w(x)$ be a homogeneous isotropic field on $\mathbb{R}^3$ with zero mean, unit variance and radial spectral measure concentrated at a point γ,

$$B(r) = \frac{\sin(\gamma r)}{\gamma r}.$$

As a numerical approximation of the field $w(x)$ we consider the following spectral model

$$w_M^{(\gamma)}(x) = M^{-1/2} \sum_{m=1}^{M} [\xi_m \cos < \lambda_m, x > + \eta_m \sin < \lambda_m, x >],$$

where ξ_m, η_m are independent standard normal variables and $< .,. >$ denotes the scalar product in $\mathbb{R}^3$. For a non-randomized model, vectors λ_m must be located "uniformly" on a hemisphere S in $\mathbb{R}^3$ with radius γ (i.e. $S \cap (-S) = \emptyset$, $S \cup (-S) = \{\|\lambda\| = \gamma\}$). For any randomized model it is necessary for vectors λ_m to be distributed according to such measures $\mu_m(d\lambda)$ that measure $\frac{1}{M} \sum_{m=1}^{M} \mu_m(d\lambda)$ is the uniform distribution on the hemisphere S. One version of the randomized model is presented below:

$$w_{I\times J}^{(\gamma)}(x)=(IJ)^{-1/2}\sum_{i=1}^{I}\sum_{j=1}^{J}[\xi_{ij}\cos<\lambda_{ij},x>+\eta_{ij}\sin<\lambda_{ij},x>],$$

$$\lambda_{ij}=(\lambda_{ij}^1,\lambda_{ij}^2,\lambda_{ij}^3),\quad t_{ij}=1-\frac{2}{I}(i-\alpha_{ij}),$$

$$\lambda_{ij}^1=\gamma t_{ij},$$

$$\lambda_{ij}^2=\gamma\sqrt{1-t_{ij}^2}\cos\left[(j-\beta_{ij})\frac{\pi}{J}\right],$$

$$\lambda_{ij}^3=\gamma\sqrt{1-t_{ij}^2}\sin\left[(j-\beta_{ij})\frac{\pi}{J}\right].$$

Here α_{ij}, β_{ij} ($i\in\{1,\dots,I\}$, $j\in\{1,\dots,J\}$) are independent random variables uniformly distributed in (0,1).

One of the methods to construct approximate models of homogeneous isotropic fields in $\mathbb{R}^3$ with an arbitrary spectral density $h(\gamma)$ consists in using spectral models of the form

$$w_{NM}(x)=\sum_{n=1}^{N}a_n w_{M_n}^{(\gamma_n)}(x,y),$$

where

a) $a_n^2=1/N$, γ_n are independent random variables identically distributed in $[0,+\infty)$ with density $h(\gamma)$ or

b) $0=b_0<b_1<\dots<b_n=+\infty$, $a_n^2=\int_{b_{n-1}}^{b_n}h(\gamma)d\gamma$, γ_n are random variables distributed in corresponding intervals $[b_{n-1},b_n)$ with density $h(\gamma)/a_n^2$.

There are some examples of correlation functions of isotropic random fields in three-dimensional space and the corresponding radial spectral densities:

1. $$h(\gamma)=\begin{cases}\frac{1}{a}, & \gamma\in(0,a),\\ 0 & \text{otherwise},\end{cases}$$

$$B(r)=\frac{Si(ar)}{ar},\quad Si(z)=\int_0^z\frac{\sin t}{t}dt;$$

2. $h(\gamma) = \dfrac{2a}{\pi(a^2+\gamma^2)}, \quad \gamma > 0, \qquad B(r) = \dfrac{1-e^{-ar}}{ar};$

3. $h(\gamma) = ae^{-a\gamma}, \quad \gamma > 0, \qquad B(r) = \dfrac{\operatorname{arctg}(r/a)}{r/a};$

4. $B(r)=e^{-ar^2}, \quad h(\gamma)=2^{-1}\pi^{-1/2}a^{-3/2}\gamma^2\exp(-\gamma^2/(4a)), \quad \gamma > 0.$

Remark. Evidently, the time of simulation for spectral models increases for larger dimensions of parameter of a random field. If values of a spectral model are calculated in nodes of a regular grid, then it is necessary to get the multitude of numbers $\cos(k\delta + \varphi)$ for many $k \in \{0, 1, 2, \ldots\}$. In this case the following relation allows to reduce considerably the time of computation

$$\cos((k+1)\delta + \varphi) = 2\cos(\delta)\cos(k\delta + \varphi) - \cos((k-1)\delta + \varphi).$$

Below we present relations between spectral measure

$$F(d\lambda) = f(\lambda)d\lambda, \quad f(\lambda) = \frac{2}{\pi}\int_0^\infty B(r)\cos(\lambda r)dr, \quad \lambda \geq 0,$$

of a stationary process which is obtained as a trace on a straight line of a homogeneous isotropic field on $\mathbb{R}^3$, and a radial spectral measure of the isotropic random field (cf., [25]):

$$F(\lambda) = \int_0^1 H(\lambda/x)dx, \quad \lambda > 0,$$

$$f(\lambda) = \int_0^1 h(\lambda/x)x^{-1}dx = \int_\lambda^\infty h(t)t^{-1}dt, \quad \lambda > 0,$$

$$h(\gamma) = -\gamma f'(\gamma), \quad \gamma > 0.$$

The last equation implies $f'(\gamma) \leq 0$.

It is easy to show that (similar to isotropic fields on a plane) if the radial spectral density h is a power function, then the density f is a power function as well with the same index.

Task 1.1. Verify the following proposition: if $g(\rho)$ denotes a radial spectral density of isotropic field on $\mathbb{R}^2$ obtained as a trace of an isotropic field on $\mathbb{R}^3$ with radial spectral density $h(\gamma)$, then

$$g(\rho) = \rho \int\limits_{\rho}^{+\infty} \frac{h(\gamma)}{\gamma(\gamma^2 - \rho^2)^{1/2}} d\gamma, \quad \rho > 0.$$

Formulate and prove the corresponding statements for isotropic fields and their traces in spaces of arbitrary dimensions.

Task 1.2. Develop methods of searching for optimal spectral models of homogeneous Gaussian fields having the minimal error for a fixed number of harmonics. Consider different ways to define the value of errors.

1.2. Technique of successive refinement of spectral models on the same probability space

1.2.1. Description of the algorithm

The technique of conditional Gaussian distributions allows to develop algorithms of successive refinement of spectral models of random processes and fields. Let us illustrate it by the following example.

Assume that for approximate simulation of a stationary Gaussian process $w(t)$ with mean zero and spectral density $f(\lambda)$, $\lambda \in [0, +\infty)$ we use a spectral model

$$w^*(x) = \sum_{j=1}^{n} \left[\xi_j \cos(\lambda_j x) + \eta_j \sin(\lambda_j x)\right], \tag{1.14}$$

where $\lambda_j \in \Lambda_j$, $\cup_{j=1}^{n} \Lambda_j = [0, +\infty)$, $\Lambda_j \cap \Lambda_k = \emptyset$ for $j \neq k$; ξ_j, η_j are independent Gaussian random variables with mean zero and variance

$$\mathbf{D}\xi_j = \mathbf{D}\eta_j = \int\limits_{\Lambda_j} f(\lambda) d\lambda.$$

To obtain better accuracy we have to increase the number of harmonics in formula (1.14). As a refined spectral model of $w(t)$ we take

$$w^{**}(x) = \sum_{j=1}^{n}\Big[\xi_j^{(1)}\cos(\lambda_j^{(1)}x) + \xi_j^{(2)}\cos(\lambda_j^{(2)}x) + $$
$$+\eta_j^{(1)}\sin(\lambda_j^{(1)}x) + \eta_j^{(2)}\sin(\lambda_j^{(2)}x)\Big],$$

where $\lambda_j^{(1)} \in \Lambda_j^{(1)}$, $\lambda_j^{(2)} \in \Lambda_j^{(2)}$, $\Lambda_j^{(1)} \cup \Lambda_j^{(2)} = \Lambda_j$, $\Lambda_j^{(1)} \cap \Lambda_j^{(2)} = \emptyset$, and $\xi_j^{(1)}$, $\eta_j^{(1)}$, $\xi_j^{(2)}$, $\eta_j^{(2)}$ are independent Gaussian variables with mean zero and variance

$$\mathbf{D}\xi_j^{(m)} = \mathbf{D}\eta_j^{(m)} = \int\limits_{\Lambda_j^{(m)}} f(\lambda)d\lambda, \quad m = 1, 2,$$

which are simulated under additional condition

$$\xi_j^{(1)} + \xi_j^{(2)} = \xi_j, \quad \eta_j^{(1)} + \eta_j^{(2)} = \eta_j \tag{1.15}$$

(see formulas (1.18) below). Thus, the refined spectral model $w^{**}(t)$ contains twice as many harmonics as $w^{*}(t)$. The same improvement procedure can be performed for model $w^{**}(t)$, etc. Condition (1.15) gives possibility to build realisations on the same probability space and to gain convergence of trajectories of spectral models to the limit process $w(t)$ (Figure 1.4).

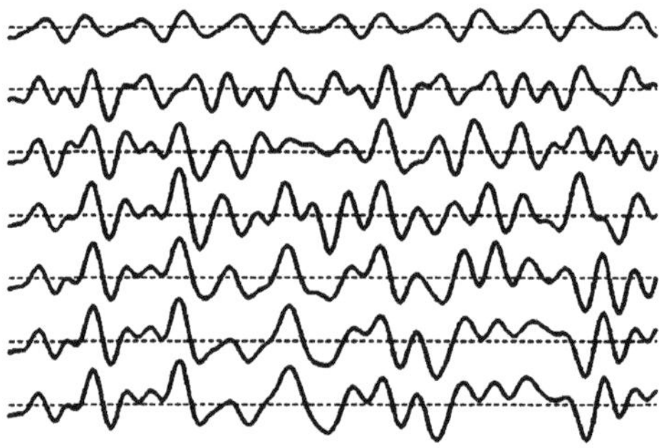

Figure 1.4. Successive refinement of spectral model of stationary Gaussian process (number of harmonics: 4, 8, 16, 32, 64, 128, 256)

1.2.2. Auxiliary statements and examples

The following statement is actual for successive refinement of spectral models on the same probability space.

Lemma 1.1. *Suppose ξ is a k-dimensional Gaussian random vector with mathematical expectation m and nonsingular correlation matrix R, A is a $l \times k$-matrix, $0 < l \leq k$, $\mathrm{rang}(A) = l$, and b is a vector of dimension l. Then conditional distribution of the vector ξ provided that $A\xi = b$, is Gaussian with mean vector*

$$m + RA^T(ARA^T)^{-1}(b - Am) \tag{1.16}$$

and correlation matrix

$$R - RA^T(ARA^T)^{-1}AR. \tag{1.17}$$

Proof. Formulas (1.16), (1.17) are true (see Appendix A) since random vector $(\xi, A\xi)^T$ is Gaussian with mathematical expectation $(m, Am)^T$ and correlation matrix

$$\begin{bmatrix} R & RA^T \\ AR & ARA^T \end{bmatrix}.$$

It remains now to ascertain that the square $l \times l$-matrix ARA^T is nonsingular:

$$\mathrm{rang}(ARA^T) = \mathrm{rang}(A\sqrt{R}\sqrt{R}A^T) = \mathrm{rang}(A\sqrt{R}) = \mathrm{rang}(A) = l.$$

Here we used the equality $\mathrm{rang}(B) = \mathrm{rang}(BB^*)$ and the fact that the rank of a rectangular matrix is invariant under multiplication by a nonsingular square matrix (see, for example, [62], [119]).

Example A. Consider two-dimensional Gaussian random vector $(\xi_1, \xi_2)^T$ with mean zero and correlation matrix

$$\begin{bmatrix} r_1 & r \\ r & r_2 \end{bmatrix}.$$

Conditional distribution of the vector $(\xi_1, \xi_2)^T$ provided $\xi_1 + \xi_2 = b$ is Gaussian with expectation

$$cb \begin{bmatrix} r_1 + r \\ r + r_2 \end{bmatrix}$$

and correlation matrix

$$c\begin{bmatrix} d & -d \\ -d & d, \end{bmatrix}$$

where $c = 1/(r_1 + 2r + r_2)$, $d = r_1 r_2 - r^2$. Hence we obtain the following formulas for numerical modelling:

$$\xi_1 = \left(\frac{r_1 r_2 - r^2}{r_1 + 2r + r_2}\right)^{1/2} \varepsilon + \frac{r_1 + r}{r_1 + 2r + r_2} b, \qquad (1.18)$$
$$\xi_2 = b - \xi_1.$$

Here ε is a standard normal random variable.

Example B. Using expressions (1.18) one can easily derive formulas for simulation of n independent Gaussian random variables $\xi_1, \ldots, \xi_n$ with mean zero and variances $\sigma_1^2, \ldots, \sigma_n^2$, provided that $\xi_1 + \ldots + \xi_n = b$:

$$\xi_1 = \left(\frac{\sigma_2^2 + \ldots + \sigma_n^2}{\sigma_1^2 + \ldots + \sigma_n^2}\right)^{1/2} \sigma_1 \varepsilon_1 + \frac{\sigma_1^2}{\sigma_1^2 + \ldots + \sigma_n^2} b,$$

$$\xi_2 = \left(\frac{\sigma_3^2 + \ldots + \sigma_n^2}{\sigma_2^2 + \ldots + \sigma_n^2}\right)^{1/2} \sigma_2 \varepsilon_2 + \frac{\sigma_2^2}{\sigma_2^2 + \ldots + \sigma_n^2} (b - \xi_1),$$

$$\xi_3 = \left(\frac{\sigma_4^2 + \ldots + \sigma_n^2}{\sigma_3^2 + \ldots + \sigma_n^2}\right)^{1/2} \sigma_3 \varepsilon_3 + \frac{\sigma_3^2}{\sigma_3^2 + \ldots + \sigma_n^2} (b - \xi_1 - \xi_2),$$

$$\ldots$$

$$\xi_{n-1} = \left(\frac{\sigma_n^2}{\sigma_{n-1}^2 + \sigma_n^2}\right)^{1/2} \sigma_{n-1} \varepsilon_{n-1} +$$
$$\frac{\sigma_{n-1}^2}{\sigma_{n-1}^2 + \sigma_n^2} (b - \xi_1 - \ldots - \xi_{n-2}),$$
$$\xi_n = b - \xi_1 - \xi_2 - \ldots - \xi_{n-1}.$$

Here $\varepsilon_1, \varepsilon_2, \ldots, \varepsilon_{n-1}$ are independent standard normal random variables.

1.3. Conditional spectral models

On the basis of the spectral decomposition, we develop here numerical algorithms that allow to model a Gaussian homogeneous random field subject to the known values of the field at fixed points (Figure 1.5). These algorithms can be used for extrapolation and interpolation of random functions.

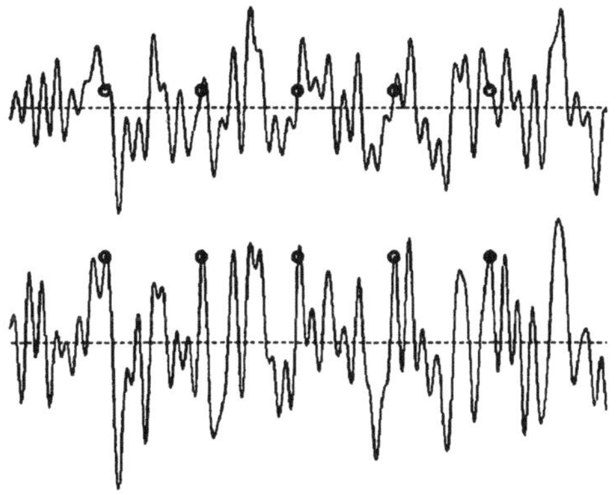

Figure 1.5. Samples of conditional spectral models of a stationary random process

1.3.1. Statement of the problem

Let $w(x)$, $x \in \mathbb{R}^k$, be a homogeneous real Gaussian random field with known expectation and spectral measure. We suppose that at the points

$$x_j \in \mathbb{R}^k, \quad j = 1, \ldots, J,$$

values of the random field are known:

$$w(x_j) = b_j, \quad j \in \{1, 2, \ldots, J\}. \tag{1.19}$$

It is necessary to construct a numerical model of the random field $w(x)$ that satisfies the condition (1.19), i.e. it is necessary to construct an algorithm for modelling the values of $w(x)$, $x \in X$, for a set $X \subset \mathbb{R}^k$. These problems arise, for example, in constructing of stochastic models of natural objects, with the experimental data being taken into account. In particular, this problem is of interest for the dynamic-probability forecast of meteorological processes (see [29], [81]).

The method proposed below is an approximate one and is based on the spectral decomposition of homogeneous random fields. Its peculiarity is that the number of the necessary arithmetic operations linearly depends on the number of elements in the set X. The error of the spectral method is substantially defined by the space dimensions of this set rather than by the number of its elements.

1.3.2. Method of solving the problem

For simplicity we assume that the mean value of the homogeneous random field $w(x)$ is equal to zero. As the numerical approximation of the field we take a spectral model:

$$w_N(x) = \sum_{n=1}^{N} c_n \left[\xi_n \cos < \lambda_n, x > + \xi_n' \sin < \lambda_n, x >\right], \tag{1.20}$$

where $c_n > 0$, $\sum_{n=1}^{N} c_n^2 < \infty$; ξ_n, ξ_n' are independent standard normal random variables; λ_n are vectors in $\mathbb{R}^k$, which are independent of the random variables ξ_n, ξ_n'; and $< .,. >$ denotes the inner product $\mathbb{R}^k$.

Thus, in order to model the random field $w(x)$ we use the spectral model (1.20). At the first stage we determine the values of c_n and simulate the vectors λ_n according to the chosen spectral model. At the second stage we simulate the independent standard normal random variables ξ_n, ξ_n' with the linear constraint

$$\sum_{n=1}^{N} d_{jn}\xi_n + d_{jn}'\xi_n' = b_j, \quad j = 1, \ldots, J, \tag{1.21}$$

where $d_{jn} = c_n \cos < \lambda_n, x_j >$, $d_{jn}' = c_n \sin < \lambda_n, x_j >$.

The following statements can be used to model the conditional Gaussian distributions.

Lemma 1.2. *Suppose that the system of linear algebraic equations $A\xi = b$ is a consistent system.*

A. *Let ξ be a Gaussian random vector with expectation m and correlation matrix R such that $b - Am \in AR(\mathbb{R}^k)$. Then the conditional distribution of the vector ξ, provided that $A\xi = b$, is the Gaussian one with expectation*

$$\begin{aligned} \mu &= m + R^{1/2}(AR^{1/2})^{+}(b - Am) = \\ &= m + RA^T(ARA^T)^{+}(b - Am) \end{aligned} \tag{1.22}$$

and correiation matrix

$$B = R - R^{1/2}(AR^{1/2})^{+}AR = R - RA^{T}(ARA^{T})^{+}AR. \tag{1.23}$$

B. *If ξ is a vector of independent standard normal random variables, formulas (1.22), (1.23) have the form*

$$\mu = A^{+}b, \tag{1.24}$$

$$B = I - A^{+}A. \tag{1.25}$$

Hereafter we denote by A^{+} the pseudoinverse of the matrix A.

We present a **proof** that is based on Lemma 1.1 and skeleton decomposition of a matrix.

Let us prove item B. Suppose that A is the $l \times k$ matrix of rank r and consider the skeleton decomposition of the matrix A:

$$A = A_1 A_2,$$

where A_1 and A_2 are matrices of rank r and dimensions $l \times r$ and $r \times k$ respectively. Random vector $\xi_1 = A_2\xi$ of dimension r satisfies the equation $A_1\xi_1 = b$. The conditions of the Lemma imply that the system $A_1\xi_1 = b$ is consistent and hence it is equivalent to the system $A_1^T A_1 \xi_1 = A_1^T b$. Matrix $A_1^T A_1$, as $r \times r$ matrix of rank r, is nonsingular and

$$\xi_1 = (A_1^T A_1)^{-1} A_1^T b = A_1^{+} b.$$

Therefore, equation

$$A\xi = b$$

is equivalent to

$$A_2\xi = A_1^{+}b,$$

and we can apply Lemma 1.1:

$$\mu = A_2^{+}A_1^{+}b = A^{+}b,$$
$$B = I - A_2^{+}A_2 = I - A^{+}A.$$

The latter equality is obvious:

$$A^+A = A_2^+A_1^+A_1A_2 = A_2^+(A_1^TA_1)^{-1}A_1^TA_1A_2 = A_2^+A_2.$$

Item B of Lemma 1.2 has been proved.

To prove Item A we present vector ξ as

$$\xi = Q\varepsilon + m,$$

where ε is a vector of independent standard normal variables, $QQ^* = R$. Condition $A\xi = A(Q\varepsilon + m) = b$ implies

$$AQ\varepsilon = b - Am. \tag{1.26}$$

This system is consistent if $b - Am \in AR(\mathbb{R}^k)$. Applying Item B of the Lemma to the Gaussian vector ε, we obtain that under condition (1.26) vector ε has the following expectation and correlation matrix

$$\mu_\varepsilon = (AQ)^+(b - Am),$$
$$B_\varepsilon = I - (AQ)^+(AQ).$$

Hence, conditional distribution of the vector $\xi = Q\varepsilon + m$ has the expectation

$$\mu = Q\mu_\varepsilon + m = m + Q(AQ)^+(b - Am)$$

and the correlation matrix

$$B = QB_\varepsilon Q^* = R - Q(AQ)^+(AQ)Q^* = R - Q(AQ)^+AR.$$

If we put $Q = R^{1/2}$ and take into account the relation (cf., for example, [1])

$$H^+ = H^T(HH^T)^+, \tag{1.27}$$

then we obtain (1.22), (1.23). Lemma 1.2 has been proved.

Lemma 1.3. *Let ξ be a vector of independent standard normal random variables. Then the conditional distribution of the vector ξ, provided that ξ is the pseudo-solution of the system of linear algebraic equations $A\xi = b$, is the Gaussian one with the expectation (1.24) and the correlation matrix (1.25).*

Lemma 1.3 is a direct consequence of the Item B in Lemma 1.2. Indeed, all the pseudo-solutions of the equation $A\xi = b$ and only they are the solutions of the consistent linear system $A^T A\xi = A^T b$. Taking into account (1.27), we get the required result.

Let us denote by ζ the vector $(\xi^T, (\xi')^T)^T$, where

$$\xi = \begin{bmatrix} \xi_1 \\ \vdots \\ \xi_N \end{bmatrix}, \quad \xi' = \begin{bmatrix} \xi'_1 \\ \vdots \\ \xi'_N \end{bmatrix}.$$

As the matrix A we take (cf. (1.21))

$$A = \begin{bmatrix} d_{11}, ..., d_{1N}, & d'_{11}, ..., d'_{1N} \\ & \vdots & \\ d_{J1}, ..., d_{JN}, & d'_{J1}, ..., d'_{JN} \end{bmatrix}$$

Then

$$M\zeta = 0, \quad M\zeta\zeta^T = I,$$

and condition (1.21) is written as

$$A\zeta = b, \quad b = \begin{bmatrix} b_1 \\ \vdots \\ b_J \end{bmatrix}.$$

According to Lemma 1.3, simulation of random variables ξ_n, ξ'_n, $n = 1, \dots, N$, in spectral model (1.20) provided

$$\|A\zeta - b\|^2 = \sum_{j=1}^{J} [w(x_j) - b_j]^2$$

is minimal, reduces to simulation of Gaussian random vector ζ with expectation

$$\mu = A^+ b$$

and correlation matrix

$$B = I - A^+ A.$$

In this case for simulation we can use the following formula

$$\zeta = \mu + G\varepsilon,$$

where ε is a vector of independent standard normal random variables and $GG^T = B$. Since B is a projection operator ($B^T = B$, $B^2 = B$) onto the subspace of dimension $2N - \text{rang}(A) = \dim(\ker(A))$, we can take $G = B$.

1.3.3. On realization of numerical algorithm

In order to calculate A^+, A^+A and $I - A^+A$ we can employ different methods, including recurrent ones (see, e. g. [1]).

As an example, we consider the algorithm for modelling the components of the vector ζ 'by steps'. We represent the vector ζ as

$$\zeta = \begin{bmatrix} \zeta_1 \\ \zeta_2 \end{bmatrix},$$

where ζ_1, ζ_2 are standard normal random vectors of smaller dimensions. We first model the vector ζ_1 provided that

$$A\zeta = A_1\zeta_1 + A_2\zeta_2 = b, \quad A = [A_1|A_2].$$

The conditional distribution of the vector ζ_1 is the Gaussian one with expectation

$$\mu = A_1^T(AA^T)^+b \tag{1.28}$$

and correlation matrix

$$B = I - A_1^T(AA^T)^+A_1. \tag{1.29}$$

It follows from Item A in Lemma 1.2.

Thus, at the first step we model the Gaussian vector ζ_1 with the expectation (1.28) and the correlation matrix (1.29):

$$\zeta_1 = \mu + Q\varepsilon, \quad QQ^T = B,$$

where ε is a vector of independent standard normal random variables. Now it remains to construct the vector ζ_2 provided that

$$A_2\zeta_2 = b - A_1\zeta_1.$$

If the dimension of the vector ζ_2 is still large, the vector ζ_2 is divided once again into two components of smaller dimension and so on.

In order to avoid the tedious procedure of factorization $B = QQ^T$ at each step, it is advisable to take the dimension of the vector ζ_1 equal to unity. In this case the algorithm to simulate the vector

$$\zeta = \begin{bmatrix} \zeta^1 \\ \vdots \\ \zeta^{2N} \end{bmatrix}$$

provided that $A\zeta = b$, can be represented as

$$\begin{aligned}
&\zeta^i = \mu_i + B_i^{1/2}\varepsilon_i, \\
&\mu_1 = a_1^T(AA^T)^+ b, \\
&\mu_i = a_i^T(A_iA_i^T)^+ \left[b - a_1\zeta^1 - a_2\zeta^2 - \ldots - a_{i-1}\zeta^{i-1}\right], \ i > 1, \\
&B_i = 1 - a_i^T(A_iA_i^T)^+ a_i.
\end{aligned}$$

Here a_i is the i-th column of the matrix A; A_i is the matrix that is derived by eliminating the first $i-1$-th columns from the matrix A; ε_i are independent standard normal random variables.

The matrices $S_i = (A_iA_i^T)^+$, $i = 2N, 2N-1, \ldots, 1$, can be calculated by the recurrence formulas (cf. [1]):

$$\begin{aligned}
S_i &= S_{i+1} + \left[\frac{1 + a_i^T S_{i+1} a_i}{(a_i^T C_i a_i)^2}\right](C_i a_i)(C_i a_i)^T - \\
&\quad - \frac{(S_{i+1}a_i)(C_ia_i)^T + (C_ia_i)(S_{i+1}a_i)^T}{a_i^T C_i a_i}, \\
C_i &= I - S_{i+1}^+ S_{i+1} \qquad (C_{2N} = I, \quad S_{2N+1} = 0),
\end{aligned}$$

if a_i is not a linear combination of the columns of A_{i+1} (i. e., $C_ia_i \neq 0$), and

$$S_i = S_{i+1} - \frac{(S_{i+1}a_i)(S_{i+1}a_i)^T}{1 + a_i^T S_{i+1} a_i}$$

otherwise.

Matrix C_i is the matrix of projection onto orthogonal complement of the linear span (a linear span will be denoted by Lo) of vectors $a_{i+1}, \ldots, a_{2N}$:

$$\begin{aligned}
C_i &= I - S_{i+1}^+ S_{i+1} = I - (A_{i+1}A_{i+1}^T)(A_{i+1}A_{i+1}^T)^+ = \\
&= I - A_{i+1}A_{i+1}^+.
\end{aligned}$$

It means that $a_i \in Lo(a_{i+1}, \dots, a_{2N})$ if and only if $C_i a_i = 0$. In case $a_i \notin Lo(a_{i+1}, \dots, a_{2N})$, the vector $C_i a_i$ is the (non-normalized) vector generated by Gram-Schmidt orthogonalization procedure for vectors $(a_{2N}, a_{2N-1}, ..., a_i)$.

Matrices C_i can be calculated recursively:

$$C_{2N} = I,$$
$$C_{i-1} = C_i,$$

for $i \neq 2N$, $a_i \in Lo(a_{i+1}, \dots, a_{2N})$, and

$$C_{i-1} = C_i - \frac{(C_i a_i)(C_i a_i)^T}{1 + a_i^T C_i a_i}$$

otherwise.

Thus, the computational scheme consists of two stages. At the first stage the matrices S_i are calculated recursively, and the following numbers are stored:

$$K_{ij} = a_i^T S_i a_j, \quad i = 2N, \dots, 1, \quad j \leq i,$$
$$K_{i0} = a_i^T S_i b.$$

At the second stage the random variables ζ^i are simulated:

$$\begin{aligned}
\mu_1 &= K_{10}, \\
\zeta^1 &= \mu_1 + (1 - K_{11})^{1/2} \varepsilon_1, \\
\mu_2 &= K_{20} - K_{21}\zeta^1, \\
\zeta^2 &= \mu_2 + (1 - K_{22})^{1/2} \varepsilon_2, \\
\mu_3 &= K_{30} - K_{31}\zeta^1 - K_{32}\zeta^2, \\
\zeta^3 &= \mu_3 + (1 - K_{33})^{1/2} \varepsilon_3, \\
&\dots
\end{aligned}$$

Task 1.3. Develop methods of joint simulation of a Gaussian stationary process and its derivatives (a Gaussian random field and its partial derivatives) on the basis of spectral models. Pay attention that it is reasonable to take different splittings of spectral space with different number of harmonics for the process and its derivatives. To construct spectral models for derivatives of fractional order $r > 0$ use the following rule of differentiation (cf. [26]):

$$\cos^{(r)}(\lambda x) = \lambda^r \cos(\lambda x + r\pi/2),$$
$$\sin^{(r)}(\lambda x) = \lambda^r \sin(\lambda x + r\pi/2).$$

Develop methods for constructing conditional joint spectral models for simulation of a Gaussian random process and its derivatives provided that some of the values of the process and the derivatives are known at a given set of points. Describe and investigate the technique of sequential improvement of the joint (conditional) spectral models on the same probability space.

1.4. Specialized models for isotropic fields on a *k*-dimensional space and on a sphere

1.4.1. Models of isotropic fields on a *k*-dimensional space

For a real-valued homogeneous isotropic random field $w(x)$ on $\mathbb{R}^k$ with zero mean the spectral decompositions for the field (1.1) and its correlation function (1.2) can be rewritten in the form [124]

$$w(x)=c_k \sum_{m=0}^{\infty} \sum_{j=1}^{h(m,k)} S_m^j(\theta_1, \dots, \theta_{k-2}, \varphi) \times$$
$$\times \int_0^{+\infty} \frac{J_{(k-2)/2+m}(\gamma r)}{(\gamma r)^{(k-2)/2}} z_m^j(d\gamma), \tag{1.30}$$

$$B(r)=\mathbf{M}w(x+y)w(x) = \int_0^{\infty} Y_k(\gamma r)\Phi(d\gamma), \quad \|y\| = r. \tag{1.31}$$

Here $(r, \theta_1, \dots, \theta_{k-1}, \varphi)$ are spherical coordinates of the point x,

$$c_k^2 = 2^{k-1}\Gamma(k/2)\pi^{k/2},$$
$$h(m,k) = (2m+k-2)(m+k-3)!/((m-2)!k!),$$

S_m^j is the orthonormal sequence of the real-valued spherical harmonics, J_ν are the Bessel functions of the first kind, z_m^j are real-valued orthogonal stochastic measures such that

$$\mathbf{M}z_m^j(d\gamma) = 0, \quad \mathbf{M}z_m^j(A)z_p^q(B) = \delta_m^p \delta_j^q \Phi(A \cap B),$$
$$Y_k(a) = 2^{(k-2)/2}\Gamma(k/2)\frac{J_{(k-2)/2}(a)}{a^{(k-2)/2}},$$

$\Phi(d\gamma)$ is a "radial" spectral measure of the homogeneous isotropic field $w(x)$.

For random fields on a plane and on the three-dimensional space we have
1) for $k = 2$

$$c_2 = \sqrt{2\pi}, \quad h(0,2) = 1, \quad h(m,2) = 2, \ m > 0,$$
$$S_0^1(\varphi) = (2\pi)^{-1/2}, \quad \varphi \in [0, 2\pi],$$
$$S_m^1(\varphi) = \pi^{-1/2}\cos(m\varphi), \quad S_m^2(\varphi) = \pi^{-1/2}\sin(m\varphi),$$
$$w(r,\varphi) = \int_0^\infty J_0(\gamma r) z_0(d\gamma) + \sqrt{2}\sum_{m=1}^\infty \Big[\cos(m\varphi)\int_0^\infty J_m(\gamma r) z_m^{(1)}(d\gamma) + $$
$$+ \sin(m\varphi)\int_0^\infty J_m(\gamma r) z_m^{(2)}(d\gamma)\Big],$$

$$B(r) = \int_0^\infty J_0(\gamma r)\Phi(d\gamma);$$

2) for $k = 3$

$$c_3 = \sqrt{2}\pi, \quad h(m,3) = 2m+1,$$
$$w(r,\theta,\varphi) = \int_0^\infty \frac{\sin(\gamma r)}{\gamma r} z_0^0(d\gamma) + \sum_{m=1}^\infty \sum_{j=0}^m A_m^j P_m^j(\cos\theta) \times$$
$$\times \Big[\cos(j\varphi)\int_0^\infty \frac{J_{m+1/2}(\gamma r)}{(\gamma r)^{1/2}} z_m^{j(1)}(d\gamma)$$
$$+ \sin(j\varphi)\int_0^\infty \frac{J_{m+1/2}(\gamma r)}{(\gamma r)^{1/2}} z_m^{j(2)}(d\gamma)\Big],$$
$$A_m^j = \left[\pi\frac{(2m+1)(m-j)!}{(m+j)!}\right]^{1/2}, \quad j > 0,$$

$$A_m^0 = \left[\frac{\pi}{2}(2m+1)\right]^{1/2},$$

$$B(r) = \int_0^\infty \frac{\sin(\gamma r)}{\gamma r}\Phi(d\gamma),$$

where P_m^j are adjoined Legendre polynomials:

$$\begin{aligned} P_m^j(x) &= (1-x^2)^{j/2}\frac{d^j}{dx^j}P_m(x) = \\ &= \frac{(1-x^2)^{j/2}}{2^m m!}\frac{d^{m+j}}{dx^{m+j}}\left[(x^2-1)^m\right], \\ P_m(x) &= \frac{1}{2^m m!}\frac{d^m}{dx^m}\left[(x^2-1)^m\right], \\ & m = 0,1,2,\ldots, \qquad j = 0,1,2,\ldots,m, \end{aligned}$$

which can be calculated on the basis of the following recursion relation [9]

$$(2m+1)xP_m^j(x) = (m-j+1)P_{m+1}^j(x) + (m+j)P_{m-1}^j(x).$$

Recall that the orthonormal sequence of real-valued harmonics S_m^j for $k=3$ (after modification of indices) can be presented as (cf., [117])

$$\begin{aligned} & S_0^0(\theta,\varphi) = \frac{1}{2\sqrt{\pi}}, \quad \theta\in(0,\pi),\ \varphi\in(0,2\pi), \\ & S_m^0(\theta,\varphi) = C_m^0 P_m(\cos\theta), \quad m = 1,2,\ldots, \\ & S_m^j(\theta,\varphi) = C_m^j P_m^j(\cos\theta)\cos(j\varphi), \ m = 1,2,\ldots,\ j = 0,1,\ldots,m, \\ & S_m^{-j}(\theta,\varphi) = C_m^j P_m^j(\cos\theta)\sin(j\varphi), \ m = 1,2,\ldots,\ j = 0,1,\ldots,m, \\ & \left[C_m^j\right]^2 = \frac{(2m+1)(m-j)!}{2\pi(m+j)!} \text{ for } j\neq 0, \quad \left[C_m^0\right]^2 = \frac{1}{4\pi}(2m+1), \end{aligned}$$

$$\int_0^{2\pi}\int_0^{\pi} S_m^j(\theta,\varphi)S_n^i(\theta,\varphi)\sin\theta\, d\theta\, d\varphi = \delta_m^n\delta_j^i,$$

while the Bessel function of a half-integer order can be expressed in terms of elementary functions:

$$J_{n+1/2}(x) = (-1)^n \Big(\frac{2}{\pi x}\Big)^{1/2} x^{n+1} \Big[\frac{1}{x}\frac{d}{dx}\Big]^n \Big[\frac{\sin x}{x}\Big],$$
$$J_{1/2}(x) = \Big(\frac{2}{\pi}\Big)^{1/2} \frac{\sin x}{\sqrt{x}},$$
$$J_{3/2}(x) = \Big(\frac{2}{\pi}\Big)^{1/2} x^{-3/2}(\sin x - x\cos x).$$

The spectral representations for isotropic fields allow one to construct effective algorithms of modelling subject to the peculiarity of isotropy (see also [36]). To approximate the infinite sum and the stochastic integral in (1.30) we can use the same principles that we used to build the spectral models in Section 1.1. In particular, the randomization principle gives an opportunity to reproduce exactly a required correlation function.

In general, a numerical model of a Gaussian homogeneous isotropic field (1.30) can be represented as

$$w^*(x) = c_k \sum_m \sum_{j=1}^{h(m,k)} S_m^j(\theta_1, \ldots, \theta_{k-2}, \varphi) \times \\ \times \sum_{i=0}^{n} a_{mi} \xi_{mi}^j \frac{J_{m+(k-2)/2}(\gamma_i r)}{(\gamma_i r)^{(k-2)/2}}. \qquad (1.32)$$

Let us study some specific features of the model (1.32) based on the representation (1.30) for isotropic fields. We consider the non-randomized model of the Gaussian homogeneous isotropic random field $w(x)$, $x = (r, \varphi) \in \mathbb{R}^2$, on the plane with correlation function $B(r) = J_0(\gamma r)$:

$$w_M(r,\varphi) = J_0(\gamma r)\xi_0 + \sqrt{2}\sum_{m=1}^{M}\Big[\cos(m\varphi)J_m(\gamma r)\xi_m^{(1)} + \\ + \sin(m\varphi)J_m(\gamma r)\xi_m^{(2)}\Big], \qquad (1.33)$$

where ξ_0, $\xi_m^{(1)}$, $\xi_m^{(2)}$ are independent standard normal random variables.

The following properties of the model (1.33) can be pointed out:

1) $\mathbf{M} w_M(r,\varphi) w_M(0,\varphi) = J_0(\gamma r)$;

2) the field $w_M(r,\varphi)$ is inhomogeneous;

3) $\mathbf{M}w_M^2(r,\varphi) = J_0^2(\gamma r) + 2\sum_{m=1}^{M} J_m^2(\gamma r) \to J_0(\gamma r)$

for $M \to +\infty$.

Thus, the approximation quality of model (1.33) worsens under increase of r, while the values of $\mathbf{M}w_M^2(r,\varphi)$ can be used to control the error.

Let us define

$$\begin{aligned} B_1(r) &= \mathbf{M}w_M(r/2,0)w_M(r/2,\pi) = \\ &= J_0^2(\gamma r/2) + 2\sum_{m=1}^{M}(-1)^m J_m^2(\gamma r), \\ \Delta_1 &= \|J_0(\gamma r) - B_1(r)\|, \end{aligned}$$

where the norm is taken in one of the following spaces $L_1[0,T]$, $L_2[0,T]$, or $C[0,T]$. To compare the error with that for model (1.12) from Section 1.1 we consider

$$\Delta_2 = \|J_0(\gamma r) - B_2(r)\|,$$

where $B_2(r) = M^{-1}\sum_{m=1}^{M}\cos(\gamma r\sin(\pi m/M))$ is the correlation function of the non-randomized version of the model (1.12) from Item 1.1.4,

$$B_2(r) \underset{M\to+\infty}{\to} J_0(\gamma r) = \pi^{-1}\int_0^{\pi}\cos(\gamma r\sin\theta)\,d\theta.$$

The values of Δ_1 and Δ_2 for $T=1$, $\gamma=20$, $M=10$ are presented in Table 1.5.

Table 1.5

Norm	L_1	L_2	C
Δ_1	0.006	0.015	0.062
Δ_2	0.032	0.078	0.330

For the same number of the summands, model (1.33) reproduces the correlation function more accurately in comparison with the non-randomized version of model (1.12). However, model (1.33) requires additional time to compute Bessel functions and its realizations are not homogeneous fields.

1.4.2. Spectral models of isotropic fields on a sphere

A random field defined on a k-dimensional sphere is called isotropic if its mathematical expectation is a constant and correlation function depends only on the angular distance between the points on the sphere.

An arbitrary isotropic random field $\xi(x)$, $\mathbf{M}\xi(x) = 0$, on a sphere $\mathbb{S}$ in $\mathbb{R}^k$ allows representation (cf. [124], [80])

$$\xi(x) = \sum_{m=0}^{\infty} \sum_{j=1}^{h(m,k)} \xi_m^j S_m^j(x), \quad x \in \mathbb{S}, \tag{1.34}$$

where ξ_m^j are random variables such that

$$\mathbf{M}\xi_m^j = 0, \quad \mathbf{M}\xi_m^j \xi_p^q = \delta_m^p \delta_j^q b_m.$$

The correlation function $B(\cos\theta) = \mathbf{M}\xi(x)\xi(y)$ of the field $\xi(x)$, where θ is the angular distance between the points x and y on the sphere, may be written in the form:

$$B(\cos\theta) = \frac{1}{\omega_k} \sum_{m=0}^{\infty} b_m \frac{C_m^{(k-2)/2}(\cos\theta)}{C_m^{(k-2)/2}(1)} h(m,k).$$

Here ω_k denotes the area of a unit sphere in $\mathbb{R}^k$,

$$\omega_k = \frac{2\pi^{k/2}}{\Gamma(k/2)};$$

C_m^j are Gegenbauer polynomials,

$$C_m^{j+1/2}(x) = \frac{2^j j!}{(2j!)} \frac{d^j}{dx^j} P_{m+j}(x),$$

$$C_m^{j+1}(x) = \frac{2^{-j}}{j!(m+j+1)} \frac{d^{j+1}}{dx^{j+1}} T_{m+j+1}(x),$$

$$j = 0, 1, 2, \ldots,$$

P_m are the Legendre polynomials and T_m are the Chebysheff polynomials. The variance of the random field $\xi(x)$ is equal to

$$\frac{1}{\omega_k}\sum_{m=0}^{\infty} b_m h(m,k).$$

The following expressions for b_m are obtained in [109]:

$$b_m = \frac{\Gamma^2\left(\frac{k-2}{2}\right)(k-2)\pi^{k/2-1}2^{k-3}m!}{\Gamma(k/2)\Gamma(m+k-2)}\times$$
$$\times\int_0^{\pi} B(\cos\theta)C_m^{(k-2)/2}(\cos\theta)(\sin\theta)^{k-2}d\theta, \quad m=0,1,2,\ldots.$$

For $k=3$ we have

$$b_m = 2\pi\int_{-1}^{-1} B(\mu)C_m^{1/2}(\mu)d\mu.$$

It is well known [124] that if an isotropic field $\xi(x)$ on a k-dimensional sphere is a restriction of a homogeneous isotropic field $w(x)$, $x \in \mathbb{R}^k$, with a radial spectral measure $\Phi(d\gamma)$, then

$$b_m = c_k^2\int_0^{\infty} J_{m+(k-2)/2}(\gamma)\gamma^{-(k-2)}\Phi(d\gamma).$$

A numerical model of isotropic field on a sphere can be constructed on the basis of decomposition (1.34). Realizations of isotropic fields on a three-dimensional sphere are presented in Figure 1.6. Similarly, on the basis of the corresponding decompositions, one can construct numerical models of vector-valued isotropic fields on a sphere and in a space.

In conclusion of the Section, let us note that models of space-time random fields, that are stationary in time and homogeneous in space, can be also constructed on the basis of spectral representation. In particular, such models were realized for simulation of undulation of the sea surface and for modelling of incompressible velocity fields (see the next Section and Chapter 2).

1.5. Certain applications of scalar spectral models

Among numerous applications of spectral models of random fields it is advisable to emphasize the simulation of stochastic structure of clouds

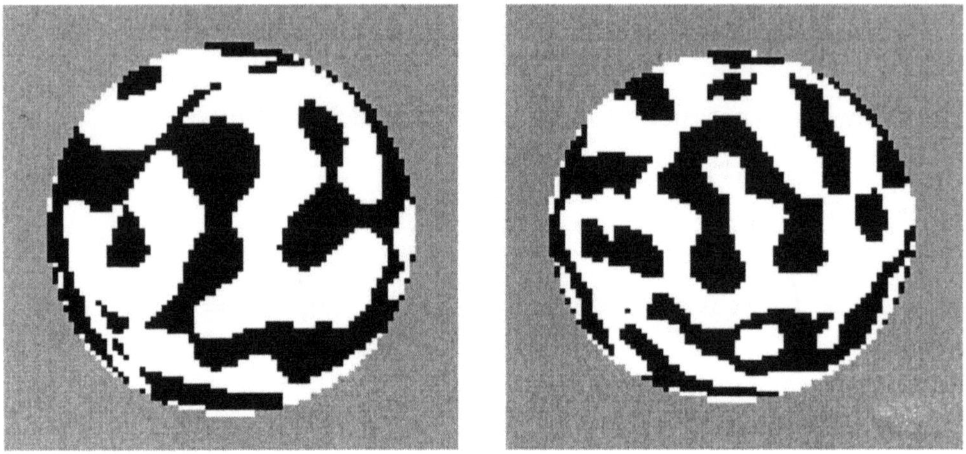

Figure 1.6. Realizations of isotropic random fields on a sphere

and undulation of the sea surface. In the Section we present a brief description of the corresponding numerical algorithms and a review of related publications.

1.5.1. Simulation of clouds

In [78] a hypothesis was put forward that cumulus clouds could be described by a stationary Gaussian process. On the basis of this hypothesis, the theoretical-experimental model of stochastic structure of cumulus cloud fields was developed (see [79], [102]). In [48] this hypothesis was used to construct a numerical model of cumulus clouds geometry for stochastic simulation of the solar radiation transfer.

Let us turn our attention to the description of the model. Assume, that cloudiness is bounded from below by the plane $z = H_0$ (the lower boundary of clouds, defined by a condensation level, slightly varies in space), and the upper random boundary of cloudiness $z = w(x, y)$ is given by the expression (**Model A**):

$$w(x,y) = H_0 + \max(v(x,y) - c, 0), \tag{1.35}$$

where $c \in (-\infty, +\infty)$, $v(x, y)$ is a homogeneous Gaussian field with mean zero, correlation function $K(x, y)$ and variance $\sigma^2 = K(0, 0)$. The cloud amount n_0 is determined by the relation

$$n_0 = 1 - \Phi(c/\sigma),$$

where Φ is the function of the standard normal distribution. A special feature of the model (1.35) is that for $c \le 0$ ($n_0 \ge 0.5$) the configuration of clouds corresponds to the structure of continuous cloudiness with breaks. Thus, in parallel with model (1.35) it is worthwile to use **Model B:**

$$w(x, y) = H_0 + \max(|v(x, y)| - c, 0), \quad c > 0. \tag{1.36}$$

In this case we have $n_0 = 2(1 - \Phi(c/\sigma))$.

By m_0 we denote the mean number of clouds per unit area. On the basis of results presented in paper [114] one can obtain

$$m_0 = c(2\pi\sigma^2)^{-3/2}(k_{20}k_{02} - k_{11}^2)^{1/2}\exp(-c^2/(2\sigma^2)), \quad c > 0$$

for model (1.35) and

$$m_0 = 2c(2\pi\sigma^2)^{-3/2}(k_{20}k_{02} - k_{11}^2)^{1/2}\exp(-c^2/(2\sigma^2))$$

for model (4). Here

$$k_{ij} = -\left.\frac{\partial^{i+j}K(x, y)}{\partial x^i \partial y^j}\right|_{x=y=0}.$$

Models (1.35), (1.36) are uniquely specified by the values c, σ and by the correlation function $K(x, y)$. Choosing these parameters appropriately one can easily tune the models to the required cloud amount and the average vertical and horizontal sizes of clouds. It enables one to use representations (1.35), (1.36) for modelling variable structures of cumulus and undulated clouds. In addition, models of stratus clouds with stochastic upper and lower boundaries can be built on the basis of homogeneous random fields.

Figure 1.7 represents examples of simulated broken clouds. Detailed information on tuning of the models and the results of computation of radiative characteristics of clouds by Monte Carlo method can be found in papers [99], [83].

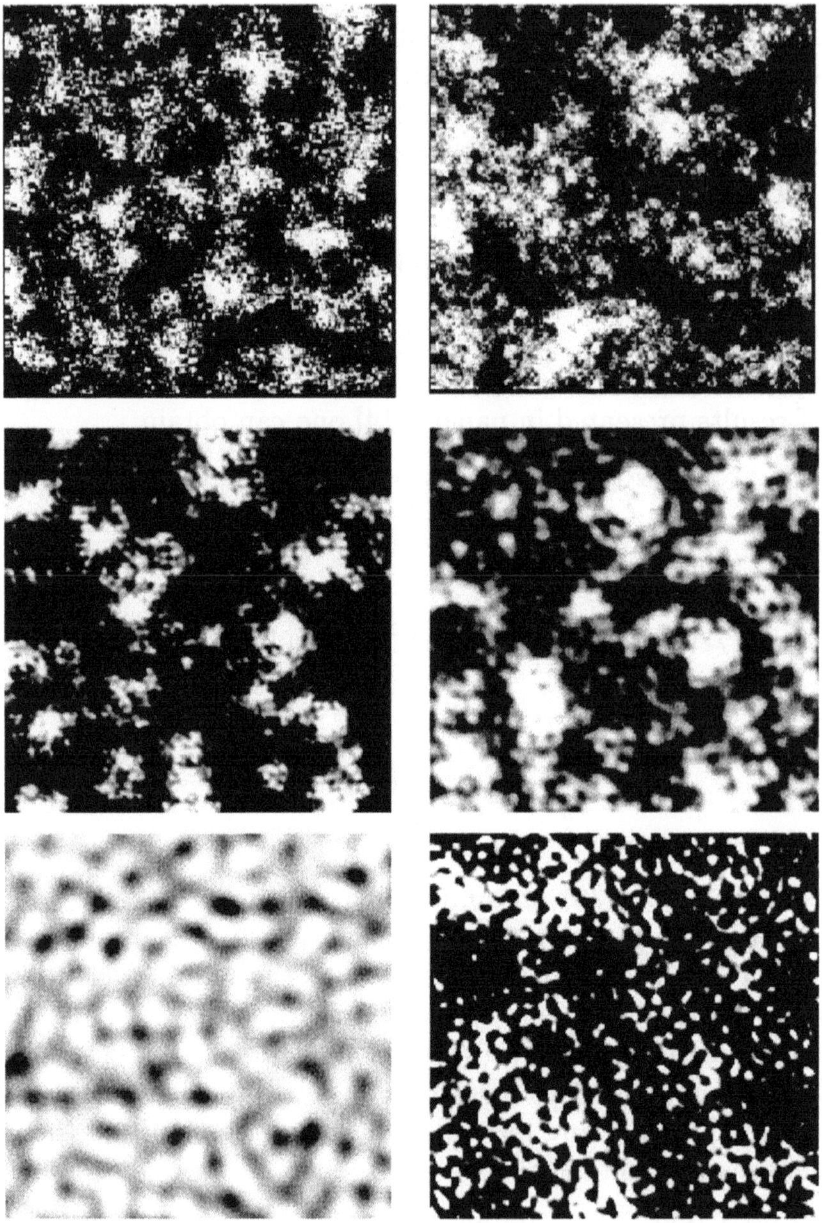

Figure 1.7. Realizations of "Gaussian models" of clouds

1.5.2. Spectral model of the sea surface undulation

A numerical model of the wind-driven undulation of the sea surface is presented below. This model has been used in papers [83], [47], [49], [8] for studying optical properties of the sea surface by Monte Carlo method. In particular, it was shown that simulation of the field of the sea surface elevations makes it possible to determine more precisely (in comparison with conventional "facet" model) the optical properties of the swell with allowance for multiple reflection effects and radiation shading by the surface elements.

The experimental data of statistical properties of wind-driven undulation of water surface indicates that it may be described with high precision by a homogeneous Gaussian random field of surface deviations from a mean level (see, e.g., [55]). Statistical properties of the sea swell are determined by the "frequency" spectrum $S(\mu)$ and the "angular" spectrum $Q(\mu,\theta)$. As approximations of the spectra we will consider the following expressions (cf. [55], [18]):

$$S(\mu)=\begin{cases} 6m_0\left(\dfrac{\mu_{\max}}{\mu}\right)^5\mu^{-1}\exp\left\{-1.2\left[\left(\dfrac{\mu_{\max}}{\mu}\right)^5-\left(\dfrac{\mu_{\max}}{\mu_1}\right)^5\right]\right\}, \\ \qquad \mu\in[0,\mu_1], \\ S(\mu_1)+(S(\mu_2)-S(\mu_1))\dfrac{\mu-\mu_1}{\mu_2-\mu_1}, \quad \mu\in(\mu_1,\mu_2), \\ 0.0078g^2\mu^{-5}, \quad \mu\in[\mu_2,\mu_3); \end{cases} \tag{1.37}$$

$$Q(\mu,\theta)=2\pi^{-1}\cos^2\theta, \quad \theta\in[-\pi/2,\pi/2). \tag{1.38}$$

Here $\mu_3\approx 30\ \text{sec}^{-1}$ is the upper boundary of the frequency spectrum of gravity waves; $\mu_{\max}$ is the frequency of the spectral maximum of the spectrum S; g is the gravity acceleration,

$$\mu_1=1.8\mu_{\max}\tilde{\mu}^{-0.7}, \quad \mu_2=2.0\mu_{\max}\tilde{\mu}^{-0.7},$$

$$\tilde{\mu}=v\mu_{\max}/g, \quad m_0=0.00127g^{-2}v^4\tilde{\mu}^{-3.19};$$

v is a wind speed (m/sec) at the height of 10 metres above the sea level. Within Model (1.37), (1.38), statistical properties of the sea undulation are determined by the wind speed v and frequency of the spectral maximum $\mu_{\max}$.

Let $w(x,y)$ denote the field of deviations of a randomly rough sea surface from a mean level (at some instant time) with the spectral characteristics (1.37), (1.38). Then the function $w(x,y)$ may be considered as a sample function of a random Gaussian homogeneous field with the spectral density in the polar coordinates ρ, θ:

$$\begin{aligned} &S_{\rho\theta}(\rho,\theta) = 2\cos^2(\theta)S_\rho(\rho)/\pi, \\ &S_\rho(\rho) = 0.5(g/\rho)^{1/2}S((g\rho)^{1/2}), \\ &\rho > 0, \quad \theta \in [-\pi/2, \pi/2]. \end{aligned} \tag{1.39}$$

The correlation function of the field $w(x,y)$ is equal to

$$\begin{aligned} &Mw(0,0)w(x,y) = \\ &= \int_0^\infty \int_{-\infty}^\infty \cos(\lambda x + \nu y) S_{\rho\theta}(\rho(\lambda,\nu), \theta(\lambda,\nu))\rho^{-1}(\lambda,\nu)\,d\nu\,d\lambda, \end{aligned}$$

where $\rho(\lambda,\nu) = (\lambda^2+\nu^2)^{1/2}$, $\theta(\lambda,\nu) = \arg(\lambda + i\nu)$.

Numerical modelling of the sea surface has been performed on the basis of the method of spectrum splitting and randomization. We present now the expression for an approximate model $w^*(x,y)$ of the Gaussian field $w(x,y)$ with spectral density (1.39). This model has been used for computations:

$$\begin{aligned} w^*(x,y) = \sum_{i=1}^m \sum_{j=1}^n a_{ij}\Big[&r_{ij}\cos(x\rho_i\cos\theta_j + y\rho_i\sin\theta_j + \varphi_{ij}) + \\ &+ r'_{ij}\cos(x\rho_i\cos\theta_j - y\rho_i\sin\theta_j + \varphi'_{ij})\Big]. \end{aligned} \tag{1.40}$$

Here ρ_i are random variables distributed over the corresponding sets A_i with probability densities proportional to S_ρ,

$$\begin{aligned} &A_i = \left[\frac{\rho^*(i-1)}{m-1}, \frac{\rho^* i}{m-1}\right), \quad i = 1,\dots,m-1, \\ &A_m = [\rho^*, \infty), \end{aligned}$$

θ_j are random variables distributed over the intervals

$$B_j = \left[\frac{\pi(j-1)}{2n}, \frac{\pi j}{2n}\right)$$

with probability density proportional to Q from (1.38), r_{ij} and r'_{ij} are random variables with Rayleigh distribution, φ_{ij} and φ'_{ij} are uniformly distributed over $[0, 2\pi]$,

$$a_{ij}^2 = \iint\limits_{A_i B_j} S_{\rho\theta}(\rho, \theta) d\theta d\rho.$$

Random variables ρ_i were simulated by the method of the inverse distribution function, and θ_j were simulated by the rejection method (with the linear majorant for $j > n/2$ and the constant majorant for $j \leq n/2$, n being even). The results of simulation (1.40) of the surface of a wind-ruffled sea are presented in Figures 1.8, 1.9.

The algorithm of modelling is defined by three parameters: m, n and ρ^* which provide the required accuracy. The field w^* has the requisite spectral density (1.39) and it is asymptotically Gaussian for

$$\max(m, n) \to \infty, \quad \rho^* \to \infty.$$

These conditions are sufficient for the weak convergence of w^* in the space of continuously differentiable functions (see Chapter 3).

It may be noted that the proposed approach to numerical modelling of a wind-ruffled sea surface may be used in solving various applied problems. In particular, a spatio-temporal model of the form

$$w_n(\vec{x}, t) = \sum_{j=1}^{n} a_j r_j \cos(< \vec{\lambda}_j, \vec{x} > + \mu_j t + \varphi_j),$$

where μ_j are connected with $\vec{\lambda}_j$ by dispersion relation, seems to be promising for solving problems in laser remote sensing of a sea surface by numerical simulation (see [8]). A realization of the spatio-temporal model of the sea swell is presented by Figure 1.10.

More detailed information on numerical models of clouds, sea swell, as well as some applications in study of radiation transfer processes, and comprehensive bibliography can be found in [96].

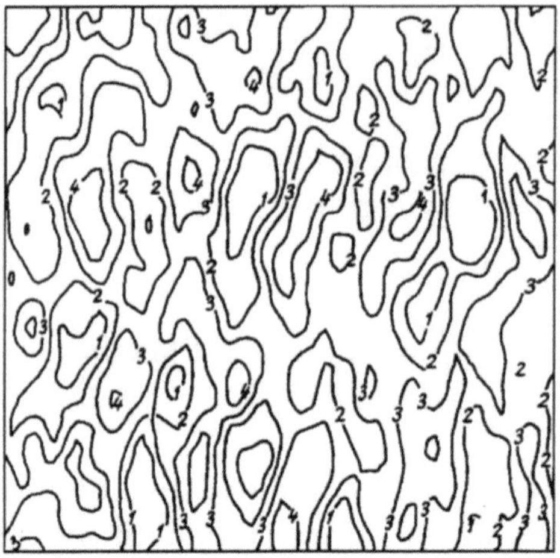

Figure 1.8. An example of the simulated topography of the sea surface roughness (spectral model)

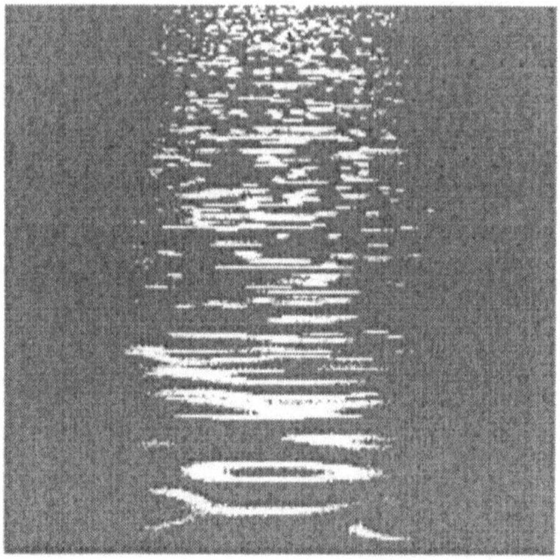

Figure 1.9. Sunglint glitter on a wind-ruffled sea (results of simulation)

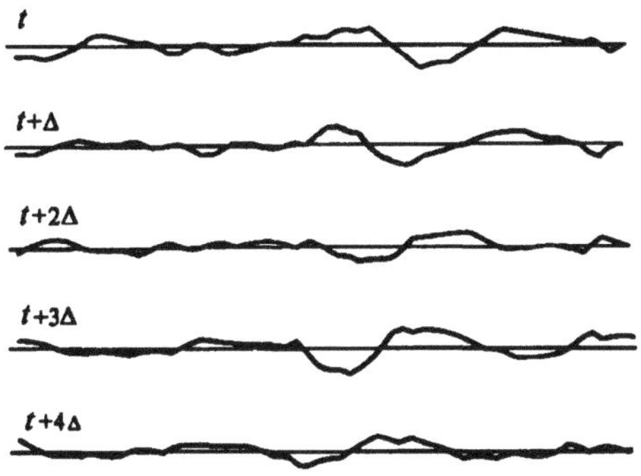

Figure 1.10. Results of simulation of the space - time structure of the sea surface (relief in a time order sequence)

1.6. Further remarks

1.6.1. Nonhomogeneous spectral models

The spectral models can be used for simulation of nonhomogeneous fields if, for example, one puts $a_j = a_j(x)$, $\lambda_j = \lambda_j(x)$ in formula (1.3). Some realizations of nonhomogeneous spectral models are shown in Figure 1.11. In general (see, for example, [31]) it is known that, if the correlation function $B(x, y)$ of a nonhomogeneous random field $w(x)$ is of the form:

$$B(t, s) = \int_{\Lambda} g(t, \lambda)\bar{g}(s, \lambda)m(d\lambda),$$

then the field $w(x)$ allows the following representation:

$$w(t) = \int_{\Lambda} g(t, \lambda)z(d\lambda), \tag{1.41}$$

where $z(d\lambda)$ is an orthogonal stochastic measure such that $m(d\lambda) = \mathbf{M}|z(d\lambda)|^2$. The principles used for constructing spectral models can be applied for the approximate modelling of nonhomogeneous random Gaussian fields of the form of (1.41).

1.6.2. Approximate modelling of Gaussian vectors of stationary type by discrete Fourier transform

The spectral method can be adapted to simulation of Gaussian vectors of the stationary type. A peculiarity of this approach is that all N

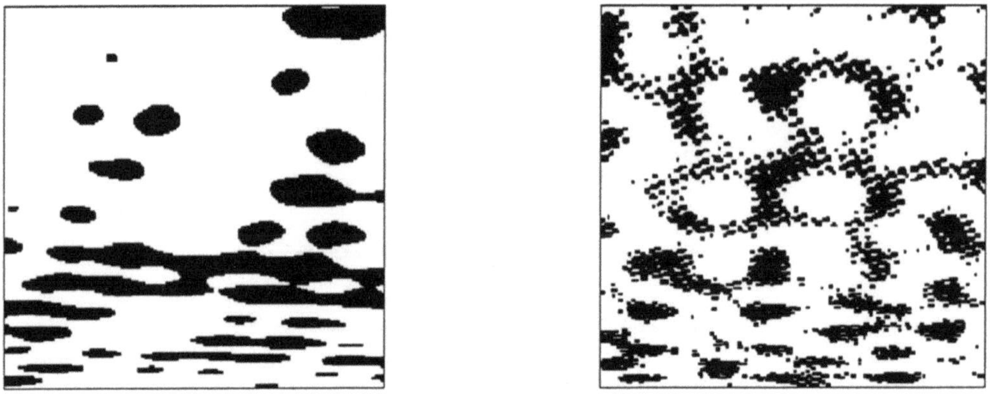

Figure 1.11. Realizations of nonhomogeneous spectral models

components of a Gaussian vector $(x_1, \dots, x_N)$ can be computed by the algorithms of the fast Fourier transform.

Below we present an interesting algorithm for simulation of Gaussian vectors of stationary type, that was reported in [11]. For even N, it was proposed to calculate the values of the Gaussian vector $(x_1, \dots, x_N)$ by formula

$$x_m = \sum_{k=0}^{N-1} c_k \exp\left(i\frac{2\pi}{N}km\right) = a_0 \operatorname{sign}(\beta_0 - 0.5) +$$

$$+(-1)^{N/2} a_{N/2} \operatorname{sign}(\beta_{N/2} - 0.5) + 2 \sum_{k=1}^{N/2-1} a_k \cos 2\pi(b_k + mk/N).$$

Here $\beta_1, \dots, \beta_{N/2}$ are independent random variables uniformly distributed in (0,1),

$$c_k = \begin{cases} a_0 \operatorname{sign}(\beta_0 - 0.5), & k = 0, \\ a_k \exp(i2\pi\beta_k), & k = 1, 2, \dots, n-1, \\ a_{N/2} \operatorname{sign}(\beta_{N/2} - 0.5), & k = N/2, \\ \bar{c}_{N-k}, & k = N/2+1, \dots, N-1, \end{cases}$$

while coefficients $a_1, \dots, a_{N/2}$ are taken from

$$a_k = \left[2\pi N^{-1} f(2\pi k/N)\right]^{1/2}, \quad k = 0, \dots, N/2.$$

It was shown in paper [11] that for the random vector $(x_1, \dots, x_N)$ the following equalities are fulfilled

$$(2\pi N)^{-1}\Big|\sum_{m=1}^{N} x_m \exp\left(i\frac{2\pi}{N}km\right)\Big|^2 = f(2\pi k/N), \qquad (1.42)$$

$$k = 0, \dots, N/2.$$

Moreover, in the case of "white noise", when

$$f(2\pi k/N) = (2\pi)^{-1}, \quad k = 0, \dots, N/2,$$

we have the equality

$$N^{-1}\sum_{m=1}^{N} x_m^2 = 1. \qquad (1.43)$$

It is asserted in [11] that «if the function $f(\lambda)$, $0 \le \lambda \le \pi$, is continuous, then it can be proved that the finite-dimensional distributions of the sequence $x_1, \dots, x_N$ for $N \to \infty$ converge in distribution to the finite-dimensional distributions of a stationary Gaussian sequence with mean zero and spectral density $f(\lambda)$, $0 \le \lambda \le \pi$". In spite of this assertion, we would hesitate to recommend the proposed method for widespread use, since the properties (1.42), (1.43) are too unnatural for Gaussian stationary sequences. In our opinion, it is more rational to use common discrete spectral models of the form

$$x_m = \sum_{k=0}^{N-1} c_k \exp\left(i\frac{2\pi}{N}km\right), \quad m = 1, \dots, N.$$

Chapter 2

Spectral models for vector-valued fields

Nowadays numerical models of vector-valued random fields are extensively used in solving applied problems, and these models have become subject of many investigations (see, for example, [53], [65], [76], [84], [85], [86] [91], [94], [107], [106], [110]). This Chapter deals with methods of numerical modelling of homogeneous vector-valued random fields based on the spectral decomposition. General relations for complex-valued and real-valued spectral models are obtained, and particular algorithms for simulation of homogeneous, isotropic, solenoidal and potential vector fields of arbitrary dimensions are presented here.

2.1. Spectral representations

2.1.1. Spectral representations for complex-valued vector random fields

Let $w(x)$ be a complex vector-valued homogeneous field,

$$w(x) = \begin{bmatrix} w_1(x) \\ \vdots \\ w_s(x) \end{bmatrix}, \quad w_r(x) \in \mathbb{C}, \ \ r = 1, \ldots, s, \ \ x \in \mathbb{R}^k$$

with mean zero and (matrix-valued) correlation function

$$\begin{aligned} K(x) &= \mathbf{M} w(x+y) w^*(y) = \\ &= \mathbf{M}\Big(\operatorname{Re} w(x+y)(\operatorname{Re} w(y))^T + \operatorname{Im} w(x+y)(\operatorname{Im} w(y))^T + \\ &\quad + i\Big\{-\operatorname{Re} w(x+y)(\operatorname{Im} w(y))^T + \operatorname{Im} w(x+y)(\operatorname{Re} w(y))^T\Big\}\Big). \end{aligned}$$

For a vector $c \in \mathbb{C}^s$ the scalar random field $c^* w(x)$ is homogeneous with correlation function $c^* K(x) c$.

If the field $w(x)$ is continuous in mean square, then [30]

$$K(x) = \int\limits_{\mathbf{R}^k} \exp(i < x, \lambda >) F(d\lambda), \tag{2.1}$$

where $F(A)$ is a matrix-valued spectral measure in $\mathbb{R}^k$, i.e., for any measurable set $A \subset \mathbb{R}^k$ the complex-valued (s,s)-matrix $F(A)$ is positive definite (hence $F(A) = F^*(A)$, $\operatorname{Re} F(A) = (\operatorname{Re} F(A))^T$, $\operatorname{Im} F(A) = -(\operatorname{Im} F(A))^T$), and for any vector $c \in \mathbb{C}^s$ the function of sets $c^* F(A) c$ is a finite measure in $\mathbb{R}^k$ (evidently, it is the spectral measure of the field $c^* w(x)$).

Some properties of a correlation function are presented below:

$$1)\ K(0) = F(\mathbb{R}^k);$$

$$2)\ K(x) = K^*(-x);$$

$$3)\ K(x)K^*(x) \le K^2(0);$$

$$4)\ (K(x) - K(y))\,(K(x) - K(y))^* \le$$
$$\le K(0)\,(2K(0) - (K(x-y) + K^*(x-y)))\,.$$

Properties 2)–4) may be obtained as a sequence of positive definiteness of correlation functions.

If measure $F(A)$ is absolutely continuous with respect to Lebesgue measure in $\mathbb{R}^k$, then

$$f(\lambda) = [f_{rt}(\lambda)] = \left[\frac{F_{rt}(d\lambda)}{d\lambda}\right]$$

is called (matrix-valued) spectral density and

$$f(\lambda) = \lim_{\varepsilon \to 0} (2\pi)^{-k} \int_{\mathbb{R}^k} \exp\left(-i < x, \lambda > -\varepsilon \|x\|^2\right) K(x) dx.$$

In particular, spectral density exists if

$$\int_{-\infty}^{+\infty} \|K(x)\| dx < +\infty,$$

and in this case

$$f(\lambda) = (2\pi)^{-k} \int_{\mathbb{R}^k} \exp(-i < x, \lambda >) K(x) dx.$$

The spectral representation of the field $w(x)$ is of the form (cf. [30])

$$w(x) = \int_{\mathbb{R}^k} \exp(i < x, \lambda >) z(d\lambda),$$

where $z(A)$ is a vector-valued spectral stochastic measure in $\mathbb{R}^k$. A spectral stochastic measure satisfies the following properties:

1) $\mathbf{M}z(A) = 0$;

2) if $A \cap B = \emptyset$, then $z(A + B) = z(A) + z(B)$;

3) $\mathbf{M}z(A)z^*(B) = F(A \cap B)$, i.e.

$$\operatorname{Re} F(A \cap B) = \mathbf{M} \operatorname{Re} z(A)(\operatorname{Re} z(B))^T + \mathbf{M} \operatorname{Im} z(A)(\operatorname{Im} z(B))^T,$$
$$\operatorname{Im} F(A \cap B) = \mathbf{M} \left(-\operatorname{Re} z(A)(\operatorname{Im} z(B))^T + \operatorname{Im} z(A)(\operatorname{Re} z(B))^T\right),$$

where A and B are measurable sets in $\mathbb{R}^k$.

Property 3 implies that if $A \cap B = \emptyset$, then

$$\mathbf{M}z(A)z^*(B) = 0. \tag{2.2}$$

2.1.2. Spectral representations for real-valued vector random fields

For the vector field $w(x)$ to be real-valued it is necessary and sufficient that $z(A) = \bar{z}(-A)$ for all measurable A and in this case

$$F(A) = \bar{F}(-A) = F^T(-A).$$

From the orthogonality property of the stochastic spectral measure $z(A)$ (cf. (2.2)) it follows that if $A \cap -A = \emptyset$, then

$$\mathbf{M}z(A)z^T(A) = 0.$$

Thus, under assumption $A \cap -A = \emptyset$, we have

$$\mathbf{M} \operatorname{Re} z(A)(\operatorname{Re} z(A))^T - \mathbf{M} \operatorname{Im} z(A)(\operatorname{Im} z(A))^T = 0,$$
$$\mathbf{M} \operatorname{Re} z(A)(\operatorname{Im} z(A))^T + \mathbf{M} \operatorname{Im} z(A)(\operatorname{Re} z(A))^T = 0$$

and, therefore,

$$\mathbf{M}\,\mathrm{Re}\,z(A)(\mathrm{Re}\,z(A))^T = \mathbf{M}\,\mathrm{Im}\,z(A)(\mathrm{Im}\,z(A))^T = \mathrm{Re}\,F(A)/2,$$
$$-\mathbf{M}\,\mathrm{Re}\,z(A)(\mathrm{Im}\,z(A))^T = \mathbf{M}\,\mathrm{Im}\,z(A)(\mathrm{Re}\,z(A))^T = \mathrm{Im}\,F(A)/2.$$

Spectral representations of the real vector-valued field $w(x)$ and its correlation function may be written in the form

$$w(x)=z\{0\}+2\int_{\mathbf{P}} \cos < x,\lambda > \mathrm{Re}\,z(d\lambda) - \sin < x,\lambda > \mathrm{Im}\,z(d\lambda) =$$
$$=\int_{\mathbf{R}^k} \cos < x,\lambda > \mathrm{Re}\,z(d\lambda) - \sin < x,\lambda > \mathrm{Im}\,z(d\lambda),$$
$$K(x)=F\{0\}+2\int_{\mathbf{P}} \cos < x,\lambda > \mathrm{Re}\,F(d\lambda) - \sin < x,\lambda > \mathrm{Im}\,F(d\lambda) =$$
$$=\int_{\mathbf{R}^k} \cos < x,\lambda > \mathrm{Re}\,F(d\lambda) - \sin < x,\lambda > \mathrm{Im}\,F(d\lambda). \tag{2.3}$$

Here $\mathbb{P}$ is a half-space of $\mathbb{R}^k$, i.e., $\mathbb{P}$ is a measurable set such that $\mathbb{P} \cap (-\mathbb{P}) = \emptyset$, $\mathbb{P} + (-\mathbb{P}) + \{0\} = \mathbb{R}^k$. For spectral density of a real vector field we have

$$f(\lambda) = (2\pi)^{-k}\Big(\int_{\mathbf{R}^k} \cos < x,\lambda > K(x)dx + i\int_{\mathbf{R}^k} \sin < x,\lambda > K(x)dx\Big).$$

Since $K(-x) = K^T(x)$ we can write

$$f(\lambda) = (2\pi)^{-k}\Big(\int_{\mathbf{P}} \cos < x,\lambda > [K(x) + K^T(x)]dx +$$
$$i\int_{\mathbf{P}} \sin < x,\lambda > [K(x) - K^T(x)]dx\Big).$$

Remark. If a random field has a real-valued correlation function, then it does not mean that the field is a real-valued one. A simple example is presented below. Let z be a complex-valued random variable such that $\mathbf{M}z = 0$, $\mathbf{M}z\bar{z} = 2A$, $\mathbf{M}zz = 0$:

$$\mathbf{M}(\operatorname{Re} z)^2 = \mathbf{M}(\operatorname{Im} z)^2 = A, \quad \mathbf{M} \operatorname{Re} z \operatorname{Im} z = 0.$$

Consider the random processes

$$\begin{aligned} u(x) &= \exp(i\lambda x)z + \exp(-i\lambda x)\bar{z} = \\ &= 2(\operatorname{Re} z \cos(\lambda x) - \operatorname{Im} z \sin(\lambda x)), \\ v(x) &= \exp(i\lambda x)z - \exp(-i\lambda x)\bar{z} = \\ &= 2i(\operatorname{Im} z \cos(\lambda x) + \operatorname{Re} z \sin(\lambda x)). \end{aligned}$$

One of the processes is purely imaginary and the other is real-valued, while both of them have the same correlation function

$$K(x) = 4A\cos(\lambda x).$$

2.2. Isotropy

A homogeneous complex vector-valued random field $w(x)$ with correlation function $K(x)$ is said to be isotropic if

$$K(x) = K(Vx) = B(\|x\|), \quad x \in \mathbb{R}^k,$$

for any orthogonal transformation V (i.e., V is a combination of rotations and reflections). For isotropic field we have $F(V^{-1}A) = F(A)$ and $K(x) = K^*(x)$.

The spectral representation (2.1) of correlation function of isotropic homogeneous field may be written in the form (cf. [37],[123])

$$B(\rho) = \int_0^\infty Y_k(\gamma\rho)G(d\gamma), \tag{2.4}$$

$$Y_k(a) = 2^{(k-2)/2}\Gamma(k/2)(a)^{-(k-2)/2}J_{(k-2)/2}(a).$$

Here J_m are Bessel functions of the first kind and $G(B) = F(\|\lambda\| \in B)$ is a matrix-valued measure in $\mathbb{R}$. Note that

$$\begin{aligned} Y_1(a) &= \cos(a), & Y_2(a) &= J_0(a), \\ Y_3(a) &= \sin(a)/a, & Y_4(a) &= 2a^{-1}J_1(a). \end{aligned}$$

If the spectral measures F and G are absolutely continuous with respect to Lebesgue measure and f, g are the corresponding spectral densities, then

$$g(\gamma) = S_k(\gamma) f(\gamma e),$$

where $S_k(\gamma) = (2\pi^{k/2}/\Gamma(k/2))\gamma^{k-1}$ is area of a sphere in $\mathbb{R}^k$ with radius γ and e is a unit vector in $\mathbb{R}^k$.

The transformations inverse to (2.4) are of the form (cf. [37], [114]):

$$G[0,\gamma) = 2^{-(k-2)/2}\Gamma^{-1}(k/2) \int_0^\infty (\gamma\rho)^{k/2} J_{k/2}(\gamma\rho)\rho^{-1} B(\rho)d\rho,$$

$$g(\gamma) = 2^{-(k-2)/2}\Gamma^{-1}(k/2) \int_0^\infty (\gamma\rho)^{k/2} J_{(k-2)/2}(\gamma\rho) B(\rho)d\rho.$$

For real-valued homogeneous isotropic vector fields the matrix-valued spectral measures G and F are real-valued:

$$F(A) = F(-A) = \bar{F}(A), \quad K(x) = K^T(x).$$

2.3. Simulation of random harmonics

2.3.1. Complex-valued harmonics

A complex-valued vector random harmonic

$$\xi(x) = \exp(i < x, \lambda >)z, \quad \xi(x), z \in \mathbb{C}^s, \quad x, \lambda \in \mathbb{R}^k, \tag{2.5}$$

where z is a complex random vector with zero mean, has correlation function

$$K(x) = \exp(i < x, \lambda >)\mathbf{M}(zz^*)$$

and its spectral measure is concentrated at the point λ:

$$\begin{aligned} F(A) &= I\{\lambda \in A\}\mathbf{M}(zz^*), \\ \operatorname{Re} F\{\lambda\} &= \mathbf{M} \operatorname{Re} z (\operatorname{Re} z)^T + \mathbf{M} \operatorname{Im} z (\operatorname{Im} z)^T, \\ \operatorname{Im} F\{\lambda\} &= -\mathbf{M} \operatorname{Re} z (\operatorname{Im} z)^T + \mathbf{M} \operatorname{Im} z (\operatorname{Re} z)^T. \end{aligned}$$

The field $\xi(x)$ is strictly homogeneous (in other words, homogeneous in narrow sense, i. e., the finite-dimensional distributions are invariant with respect to shifts) if and only if the random vector z has the structure

$$z = \exp(i\theta)z_0,$$

where z_0 is an arbitrary complex-valued random vector, while θ is a random variable independent of z_0 and uniformly distributed in $[0, 2\pi]$. Obviously, we have

$$\mathbf{M}zz^* = \mathbf{M}z_0 z_0^*.$$

A harmonic with a random frequency λ and amplitude z_λ dependent on the frequency

$$\begin{aligned} &\xi(x) = \exp(i < x, \lambda >)z_\lambda, \\ &\mathbf{M}z_\lambda = 0, \quad \mathbf{M}z_\lambda z_\lambda^* = F(d\lambda)/\mu(d\lambda), \end{aligned} \tag{2.6}$$

where λ is a random vector distributed in $\mathbb{R}^k$ according to probability measure μ, absolutely continuous with respect to F, has the correlation function

$$K(x) = \int_{\mathbb{R}^k} \exp(i < x, \lambda >)F(d\lambda)$$

and the spectral measure $F(d\lambda)$.

Remark 1. Basically in the capacity of the measure μ for harmonic (2.6) one may take an arbitrary probability measure in $\mathbb{R}^k$ absolutely continuous with respect to F. The authors of paper [107] propose

$$\mu(d\lambda) = \operatorname{tr} F(d\lambda)/\operatorname{tr} F(\mathbb{R}^k),$$

where tr denotes the trace of a matrix. The value $\operatorname{tr} F(\mathbb{R}^k)$ may be interpreted as the "full energy" of the field and $\operatorname{tr} F(d\lambda)$ as the "energy" of the frequencies $d\lambda$.

Remark 2. If $z_\lambda = z$ does not depend on λ, then

$$K(x) = \int_{\mathbb{R}^k} \exp(i < x, \lambda >)\mu(d\lambda)\mathbf{M}[zz^*].$$

2.3.2. Real-valued harmonics

Let us consider a real-valued case. Vector harmonic

$$\begin{aligned}\xi(x) &= \exp(i < x, \lambda >)z + \exp(-i < x, \lambda >)\bar{z} = \\ &= 2\left(\cos < x, \lambda > \operatorname{Re} z - \sin < x, \lambda > \operatorname{Im} z\right), \end{aligned} \tag{2.7}$$

where $\mathbf{M}z = 0$, $\mathbf{M}zz^T = 0$, has correlation function

$$K(x) = 2\cos < x, \lambda > \operatorname{Re}\mathbf{M}[zz^*] - 2\sin < x, \lambda > \operatorname{Im}\mathbf{M}[zz^*]$$

and spectral measure being concentrated at the points λ, $-\lambda$,

$$F\{\lambda\} = \mathbf{M}[zz^*] = \bar{F}\{-\lambda\}.$$

The equality $\mathbf{M}zz^T = 0$ means

$$\begin{aligned}&\mathbf{M}\operatorname{Re} z(\operatorname{Re} z)^T - \mathbf{M}\operatorname{Im} z(\operatorname{Im} z)^T = 0,\\ &\mathbf{M}\operatorname{Re} z(\operatorname{Im} z)^T + \mathbf{M}\operatorname{Im} z(\operatorname{Re} z)^T = 0\end{aligned}$$

and, hence,

$$\begin{aligned}&\mathbf{M}\operatorname{Re} z(\operatorname{Re} z)^T = \mathbf{M}\operatorname{Im} z(\operatorname{Im} z)^T = \operatorname{Re} F\{\lambda\}/2,\\ &-\mathbf{M}\operatorname{Re} z(\operatorname{Im} z)^T = \mathbf{M}\operatorname{Im} z(\operatorname{Re} z)^T = \operatorname{Im} F\{\lambda\}/2.\end{aligned}$$

If

$$z = \begin{bmatrix} z_1 \\ \vdots \\ z_s \end{bmatrix} = \begin{bmatrix} \rho_1 \exp(i\varphi_1) \\ \vdots \\ \rho_s \exp(i\varphi_s) \end{bmatrix}, \quad \xi(x) = \begin{bmatrix} \xi_1(x) \\ \vdots \\ \xi_s(x) \end{bmatrix},$$

then (2.7) may be rewritten in the form

$$\xi_m(x) = 2\rho_m \cos\left(< x, \lambda > + \varphi_m\right), \quad m = 1, \ldots, s. \tag{2.8}$$

Formula (2.8) proves to be more effective for numerical modelling than (2.7). A strictly homogeneous harmonic may be presented in the form

$$\xi_m(x) = 2\rho_m \cos(< x, \lambda > + \varphi_m^0 + \theta), \quad m = 1, \ldots, s,$$

where the value of θ is independent of ρ_m and φ_m^0 is uniformly distributed in $[0, 2\pi]$.

The harmonic

$$\xi(x) = \cos < x, \lambda > \operatorname{Re} z(\lambda) - \sin < x, \lambda > \operatorname{Im} z(\lambda) = \\ = \begin{bmatrix} \rho_1(\lambda) \cos(< x, \lambda > + \varphi_1(\lambda)) \\ \vdots \\ \rho_s(\lambda) \cos(< x, \lambda > + \varphi_s(\lambda)) \end{bmatrix}, \tag{2.9}$$

$$z(\lambda) = \begin{bmatrix} z_1(\lambda) \\ \vdots \\ z_s(\lambda) \end{bmatrix} = \begin{bmatrix} \rho_1(\lambda) \exp(i\varphi_1(\lambda)) \\ \vdots \\ \rho_s(\lambda) \exp(i\varphi_s(\lambda)) \end{bmatrix},$$

$$\mathbf{M} z(\lambda) z^*(\lambda) = \begin{cases} 2F(d\lambda)/\mu(d\lambda), & \lambda \neq 0, \\ F\{0\}/\mu\{0\}, & \lambda = 0, \end{cases}$$

$$\mathbf{M} z(\lambda) = 0, \quad \mathbf{M} z(\lambda) z^T(\lambda) = 0,$$

where λ is a random variable in $\mathbb{R}^k$ distributed according to symmetric ($\mu(d\lambda) = \mu(-d\lambda)$) probability measure μ (absolutely continuous with respect to F) has the correlation function (2.3) and the spectral measure $F(d\lambda)$. Moreover, the following is fulfilled:

$$\mathbf{M} \operatorname{Re} z(\lambda) (\operatorname{Re} z(\lambda))^T = \mathbf{M} \operatorname{Im} z(\lambda) (\operatorname{Im} z(\lambda))^T = \\ = \operatorname{Re} F\{d\lambda\}/\mu(d\lambda),$$

$$-\mathbf{M} \operatorname{Re} z(\lambda) (\operatorname{Im} z(\lambda))^T = \mathbf{M} \operatorname{Im} z(\lambda) (\operatorname{Re} z(\lambda))^T = \\ = \operatorname{Im} F\{d\lambda\}/\mu(d\lambda),$$

$$\mathbf{M} \begin{bmatrix} \operatorname{Re} z(\lambda) \\ \operatorname{Im} z(\lambda) \end{bmatrix} \begin{bmatrix} \operatorname{Re} z(\lambda) \\ \operatorname{Im} z(\lambda) \end{bmatrix}^T = \\ = \frac{1}{\mu(d\lambda)} \begin{bmatrix} \operatorname{Re} F\{d\lambda\} & -\operatorname{Im} F\{d\lambda\} \\ \operatorname{Im} F\{d\lambda\} & \operatorname{Re} F\{d\lambda\} \end{bmatrix}. \tag{2.10}$$

If $F\{0\} = 0$, then in the capacity of the measure μ one may take a measure concentrated on a half-space. In this case

$$\mathbf{M} z(\lambda) z^*(\lambda) = 4F(d\lambda)/\mu(d\lambda),$$

and in the right-hand side of (2.10) coefficient 2 will appear.

Remark. Sometimes for the real-valued case it is more convenient to use spectral representations in the following form (cf. (2.3))

$$w(x) = \int_{\mathbf{P}} \cos < x, \lambda > z_1(d\lambda) - \sin < x, \lambda > z_2(d\lambda);$$

$$K(x) = \int_{\mathbf{P}} \cos < x, \lambda > F_1(d\lambda) - \sin < x, \lambda > F_2(d\lambda);$$

$$\mathbf{M} z_i(d\lambda) z_i^T(d\lambda) = F_1(d\lambda), \quad i = 1, 2;$$

$$-\mathbf{M} z_1(d\lambda) z_2^T(d\lambda) = \mathbf{M} z_2(d\lambda) z_1^T(d\lambda) = F_2(d\lambda);$$

$$F_1(d\lambda) = 2 \operatorname{Re} F(d\lambda), \quad F_2(d\lambda) = 2 \operatorname{Im} F(d\lambda), \quad \lambda > 0;$$

where F_1 and F_2 are corresponding symmetric and skew-symmetric matrix-valued measures on the half-space $\mathbb{P}$. In this context the harmonic (2.9) can be written in the form

$$\xi(x) = Z_1(\lambda) \cos < x, \lambda > - Z_2(\lambda) \sin < x, \lambda >,$$

where λ is a vector distributed according the measure μ on the half-space $\mathbb{P}$,

$$\mathbf{M} Z_i(\lambda) Z_i^T(\lambda) = F_1(d\lambda)/\mu(d\lambda), \quad i = 1, 2;$$

$$-\mathbf{M} Z_1(\lambda) Z_1^T(\lambda) = \mathbf{M} Z_2(\lambda) Z_2^T(\lambda) = F_2(d\lambda)/\mu(d\lambda).$$

2.3.3. About simulation of complex-valued Gaussian vectors

Modelling of a complex-valued random vector z with zero mean and specified covariance matrix $F = \mathbf{M} z z^*$ (with additional condition $\mathbf{M} z z^T = 0$ for real-valued harmonics) is the basic problem for constructing homogeneous vector-valued harmonics (2.5)–(2.7), (2.9). A conventional solution of this problem consists of two stages: modelling of an orthonormal vector ε, $\mathbf{M} \varepsilon \varepsilon^* = E$ and subsequent linear transformation

$$z = A\varepsilon, \quad \text{where } AA^* = F. \tag{2.11}$$

Remark. Correlation matrix $\mathbf{M} y y^*$ does not define uniquely correlation coefficients for all components $\operatorname{Re} y_i$, $\operatorname{Im} y_i$ of the complex-valued random vector y. Therefore, usually it is supposed (often without mentioning) that

$$\mathbf{M} y y^T = 0.$$

In particular, for an orthonormal complex vector ε, $\mathbf{M}\varepsilon\varepsilon^* = E$, equation $\mathbf{M}\varepsilon\varepsilon^T = 0$ means that real and imaginary parts of the vector have the same correlation matrix $E/2$ and they are orthogonal to each other:

$$\mathbf{M}\,\mathrm{Re}\,\varepsilon(\mathrm{Re}\,\varepsilon)^T = \mathbf{M}\,\mathrm{Im}\,\varepsilon(\mathrm{Im}\,\varepsilon)^T = E/2,$$
$$\mathbf{M}\,\mathrm{Re}\,\varepsilon(\mathrm{Im}\,\varepsilon)^T = 0.$$

Let us assume that the positive definite matrix F is not singular. The linear transformation A in (2.11) is uniquely defined up to a unitary operator U: vectors $A\varepsilon$ and $AU\varepsilon$ have the same covariance matrix F.

If we require that matrix A is self-conjugated, $A = A^*$, then $A = F^{1/2}$. In this case matrix A may be found by the following recurrent procedure:

$$A_0 = 0, \quad A_{n+1} = A_n + \left(2 \parallel F \parallel^{1/2}\right)^{-1} (F - A_n^2)$$

(see, e.g., [104]), or by the formula $A = W(D)^{1/2}W^*$, where D is a diagonal matrix with eigenvalues of the matrix F as diagonal elements, and W is a matrix, whose columns are eigenvectors of the matrix F, $F = WDW^*$.

If we require that matrix A be a lower triangular one, then it gives us the well-known recurrent formulas for the matrix elements A_{rt}

$$A_{11} = (F_{11})^{1/2}\exp(i\varphi_1),$$
$$A_{rt} = \left(F_{rt} - \sum_{k=1}^{t-1} A_{rk}\bar{A}_{tk}\right) / A_{tt}, \quad t = 1, \dots, r-1,$$
$$A_{rr} = \left(F_{rr} - \sum_{t=1}^{r-1} |A_{rt}|^2\right)^{1/2} \exp(i\varphi_r), \quad r = 2, \dots, s$$

(the diagonal elements of the matrix F are real), where $\varphi_1, \dots, \varphi_s$ are arbitrary real constants. The ambiguity of constructing the matrix A is explained by the fact that matrix

$$A \times \begin{bmatrix} \exp(i\varphi_1) & 0 & \dots & 0 \\ 0 & \exp(i\varphi_2) & \dots & 0 \\ \vdots & & \ddots & \vdots \\ 0 & \dots & & \exp(i\varphi_s) \end{bmatrix}$$

is also a lower triangular one.

2.4. Spectral models of homogeneous Gaussian vector-valued fields

Let $w(x)$ be a Gaussian homogeneous vector field with mean zero, correlation function $K(x)$ and spectral measure $F(d\lambda)$. An approximate model $w_n(x)$ of the form

$$w_n(x) = \xi_1(x) + \ldots + \xi_n(x),$$

where $\xi_i(x)$ are random harmonics considered in the previous section, will be called a *spectral model.* We assume that

$$\mathbf{M}\xi_i(x)\xi_j^*(y) = 0, \quad \text{for } i \neq j, \quad x, y \in \mathbb{R}^k.$$

Then

$$K_{(n)}(x) = \sum_{j=1}^{n} K_j(x), \quad F_{(n)}(x) = \sum_{j=1}^{n} F_j(x).$$

Here $K_{(n)}$, $F_{(n)}$, K_j, F_j are the corresponding correlation functions and spectral measures of the fields w_n and ξ_j.

One may use a lot of various procedures to construct spectral models $w_n(x)$, which ensure convergence to the Gaussian vector field $w(x)$, as $n \to \infty$ (such procedures were studied for scalar fields in Chapter 1). Let us present two typical models:

$$(1) \quad F_j(d\lambda) = F(d\lambda)/n, \quad j = 1, \ldots, n;$$

$$(2) \quad \mathbb{R}^k = \sum_{j=1}^{n} \Lambda_j, \quad F_j(d\lambda) = \begin{cases} F(d\lambda), & d\lambda \subset \Lambda_j, \\ 0, & d\lambda \subset \mathbb{R}^k - \Lambda_j. \end{cases}$$

Note that the fields w_n are asymptotically Gaussian as $n \to \infty$ for model (1), while for model (2) the following condition is sufficient for convergence:

$$\max_{j \leq n} F(\Lambda_j) \to 0, \quad \text{as } n \to \infty.$$

Obviously, for both models the spectral measures $F_{(n)}(d\lambda)$ of the approximate fields $w_n(x)$ coincide with the spectral measure $F(d\lambda)$ of the limit field $w(x)$. This is attained by randomized choice of the harmonic frequencies.

2.5. Examples of simulation

2.5.1. Gradient of isotropic scalar field

Let $v(x)$ be a homogeneous, scalar, mean square differentiable complex-valued random field on $\mathbb{R}^k$ with correlation function $R(x)$ and spectral measure $\mu(d\lambda)$. Potential vector field

$$w(x) = \operatorname{grad} v(x) = \left(\frac{\partial}{\partial x_1}v(x), \dots, \frac{\partial}{\partial x_k}v(x)\right)^T \tag{2.12}$$

is homogeneous with correlation function

$$K(x) = \left[-\frac{\partial R(x)}{\partial x_r \partial x_t}\right]_{r,t=1,\dots,k}$$

and spectral measure

$$F(d\lambda) = [\lambda_r \lambda_t \mu(d\lambda)]_{r,t=1,\dots,k} \,.$$

If the field $v(x)$ is isotropic,

$$R(x) = B(\| x \|), \quad \mu(d\lambda) = S_k^{-1}(\gamma) d\sigma(\gamma) \nu(d\lambda),$$

where $\gamma = \|\lambda\|$, $S_k(\gamma)$ is area of a sphere in $\mathbb{R}^k$ with radius γ, $d\sigma(\gamma)$ is an area element of this sphere, then the field (2.12) is isotropic in the following sense [125]:

$$\mathbf{M}w(x) = 0, \quad \mathbf{M}w(x)w^*(y) = V\left[\mathbf{M}w(Vx)w^*(Vy)\right]V^*, \tag{2.13}$$

where V is an arbitrary orthogonal transformation in $\mathbb{R}^k$.

2.5.2. Solenoidal and potential isotropic random fields

Henceforth fields which are isotropic in the sense (2.13) will be called bi-isotropic in order to distinguish this concept from the definition in Section 2.2. So, the homogeneous vector field $w(x) \in \mathbb{R}^k$, $x \in \mathbb{R}^k$, will be called bi-isotropic if the first and the second moments of the fields $w(x)$ and $Vw(Vx)$ coincide for any orthogonal transformation V. It means that

$$K(x) = V\ K(Vx)\ V^*,$$

where $K(x)$ is the correlation function of the field $w(x)$.

The spectral measure of bi-isotropic field is of the form [125]

$$F(d\lambda) = S_k^{-1}(\gamma)d\sigma(\gamma)\left[L(\lambda)\varphi(d\lambda) + (E - L(\lambda))\psi(d\lambda)\right], \qquad (2.14)$$

where $\gamma = \|\lambda\|$, $S_k(\gamma)$ is area of a sphere in $\mathbb{R}^k$ with radius γ, $d\sigma(\gamma)$ is an area element of this sphere, E is the $k \times k$ identity matrix, $L(\lambda) = \lambda\lambda^T/\|\lambda\|^2$, φ and ψ are some finite measures on $[0, \infty)$. It is easy to verify that

$$\mathbf{M}\|w\|^2 = \int_{\mathbf{R}^k} \operatorname{tr} F(d\lambda) = \int_0^\infty \varphi(d\gamma) + (k-1)\int_0^\infty \psi(d\gamma).$$

For potential bi-isotropic field (2.12) the following is fulfilled:

$$\varphi(d\gamma) = \gamma^2\nu(d\gamma), \quad \psi(d\gamma) = 0.$$

In general case a homogeneous potential bi-isotropic vector field is gradient of a locally homogeneous and locally isotropic scalar random field (see [125]).

If $w(x)$ is a bi-isotropic field with spectral measure (2.14), then the scalar field $\operatorname{div} w(x)$ is isotropic with "radial" spectral measure $\gamma^2\varphi(d\gamma)$. So, the field $w(x)$ is solenoidal ($\operatorname{div} w(x) = 0$) if and only if $\varphi(d\gamma) = 0$.

Finally, we notice that coincidence of measures $\varphi(d\gamma) = \psi(d\gamma)$ corresponds to a homogeneous random field which is isotropic and bi-isotropic simultaneously. Such a field consists of uncorrelated components.

A general representation of correlation functions for bi-isotropic random fields is obtained in [125].

2.5.3. Vector-valued isotropic fields on plane and in space

In two-dimensional case any correlation function of a potential bi-isotropic field (2.12) may be written in the form

$$K(x) = \int_0^\infty \left[-J_2(\gamma\rho)X + J_1(\gamma\rho)(\gamma\rho)^{-1}E\right]\varphi(d\gamma),$$

where $\rho = \|x\|$, $X = xx^*/\|x\|^2$ and E is the 2×2 identity matrix.

Realizations of spectral models of a scalar isotropic field on the plane with correlation function $J_0(c\|x\|)$ and the gradient of this field are presented in Figure 2.1. This case corresponds to the spectral measures ν, φ concentrated at the point $\gamma = c$. More general models may be obtained by summation of such fields of different scale (see Figure 2.2).

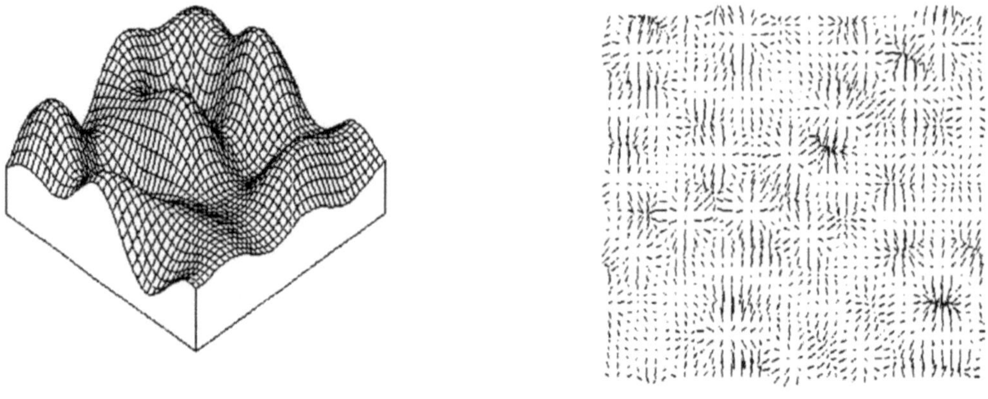

Figure 2.1. Realizations of spectral models of a scalar isotropic field with correlation function $J_0(cx)$ and its gradient

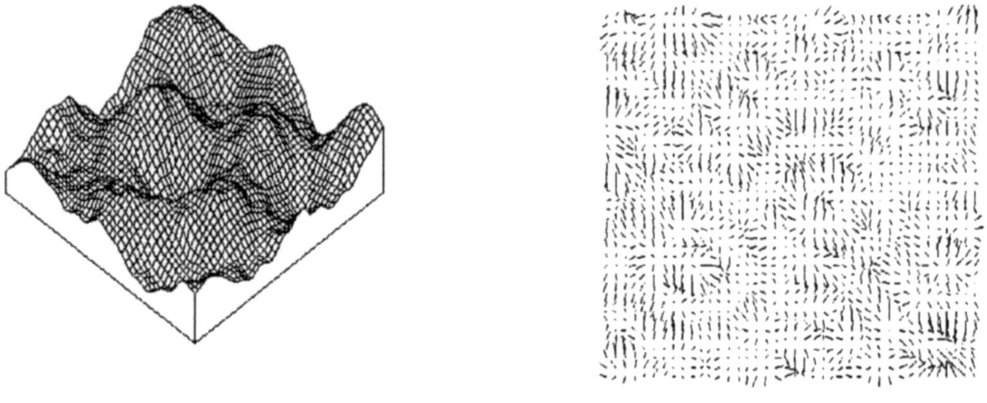

Figure 2.2. Realizations of a scalar isotropic field and its gradient (spectral models)

A correlation function of any solenoidal bi-isotropic two-dimensional vector field may be written in the form

$$K(x) = \int_0^\infty \left\{ J_2(\gamma\rho)X + \left[J_0(\gamma\rho) - J_1(\gamma\rho)(\gamma\rho)^{-1}\right] E \right\} \psi(d\gamma).$$

Some realizations of spectral models of solenoidal fields on the plane are presented in Figure 2.3.

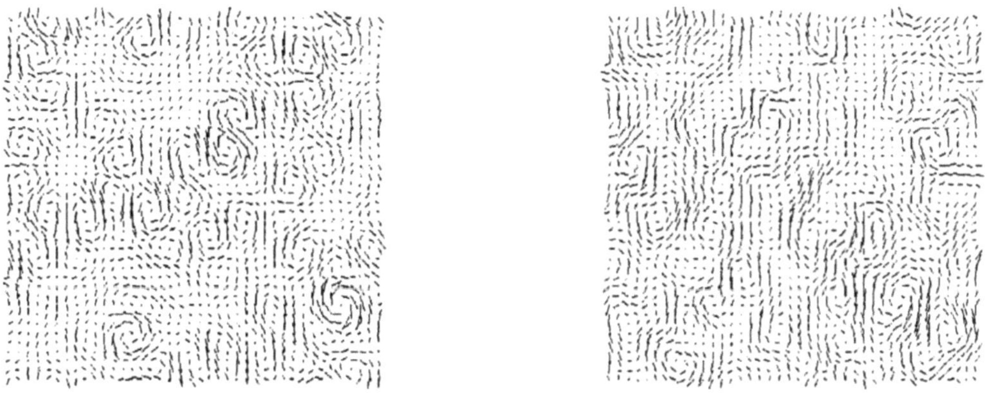

Figure 2.3. Realizations of spectral models of solenoidal bi-isotropic random fields on the plane

Below we present formulas which enables us to construct spectral models of real bi-isotropic fields for any dimension.

First, we describe an algorithm for numerical modelling of bi-isotropic solenoidal vector field $w(x)$ on the plain ($x \in \mathbb{R}^2$, $w(x) \in \mathbb{R}^2$) with mean zero, "variance" unit, $M\|w\|^2 = 1$, and measure $\psi(d\gamma)$ concentrated at point γ, $\psi(\{\gamma\}) = 1$. For $k = 2$ we have

$$L(\lambda) = \frac{1}{\|\lambda\|^2}\begin{bmatrix} \lambda_1^2 & \lambda_1\lambda_2 \\ \lambda_1\lambda_2 & \lambda_2^2 \end{bmatrix},$$

$$E - L(\lambda) = \frac{1}{\|\lambda\|^2}\begin{bmatrix} 1-\lambda_1^2 & -\lambda_1\lambda_2 \\ -\lambda_1\lambda_2 & 1-\lambda_2^2 \end{bmatrix}.$$

If $\lambda_1 = \gamma\cos\omega$, $\lambda_2 = \gamma\sin\omega$, then

$$L(\lambda) = \frac{1}{\|\lambda\|^2}\begin{bmatrix} \cos^2\omega & \cos\omega\sin\omega \\ \cos\omega\sin\omega & \sin^2\omega \end{bmatrix},$$

$$E - L(\lambda) = \frac{1}{\|\lambda\|^2}\begin{bmatrix} \sin^2\omega & -\cos\omega\sin\omega \\ -\cos\omega\sin\omega & \cos^2\omega \end{bmatrix}.$$

Since in the case under consideration the matrix-valued spectral measure $F(d\gamma)$ is concentrated on the circle with radius γ, we can use the following spectral model

$$w_n^{(\gamma)}(x) = n^{-1/2} \sum_{j=1}^{n} \xi_j(x), \tag{2.15}$$

$$\xi_j(x) = z_j \cos < x, \lambda_j > -z_j' \sin < x, \lambda_j >,$$

$$\lambda_j = (\gamma \cos \omega_j, \gamma \sin \omega_j), \quad \omega_j = (j - \alpha)\pi/n \in (0, \pi),$$

where α is a random variable uniformly distributed in (0,1) and, since $-Mz_j(z_j')^T = Mz_j'(z_j)^T = 2 \operatorname{Im} F(d\lambda) = 0$, z_j and z_j' are independent two-dimensional Gaussian vectors such that

$$Mz_j(z_j)^T = Mz_j'(z_j')^T = E - L(\lambda)$$

(turn our attention to the fact that vectors λ_j are simulated on a semicircle). It is easy to verify that random vectors z_j and z_j' can be simulated as follows:

$$z_j = \begin{bmatrix} \sin \omega_j \\ -\cos \omega_j \end{bmatrix} \eta_j, \quad z_j' = \begin{bmatrix} \sin \omega_j \\ -\cos \omega_j \end{bmatrix} \eta_j', \tag{2.16}$$

where η_j, η_j' are standard normal variables. It gives the resultant formula for simulation

$$\begin{aligned} \xi_j(x, y) &= \begin{bmatrix} \sin \omega_j \\ -\cos \omega_j \end{bmatrix} \Big\{ \eta_j \cos\left(\gamma[x \cos \omega_j + y \sin \omega_j]\right) - \\ &\quad -\eta_j' \sin\left(\gamma[x \cos \omega_j + y \sin \omega_j]\right) \Big\} = \qquad (2.17) \\ &= \begin{bmatrix} \sin \omega_j \\ -\cos \omega_j \end{bmatrix} \sqrt{-2 \ln \alpha_j} \cos\left(\gamma[x \cos \omega_j + y \sin \omega_j] + 2\pi\beta_j\right), \end{aligned}$$

where α_j, β_j are independent random variables uniformly distributed in (0,1).

An approximate model of bi-isotropic solenoidal field with an arbitrary measure ψ can be constructed as a superposition of models (2.15),

$$w^*(x) = \sum_{\gamma} a(\gamma) w_{n_\gamma}^{(\gamma)}(x),$$

with use of standard technique of partitioning and randomization with respect to measure ψ. More general models can be built by partitioning of the whole spectral half-space and subsequent randomization of vectors λ in corresponding regions (see formulas for simulation of scalar isotropic fields in Chapter 1).

If we replace vector $\begin{bmatrix} \sin\omega_j \\ -\cos\omega_j \end{bmatrix}$ by $\begin{bmatrix} \cos\omega_j \\ \sin\omega_j \end{bmatrix}$ in expressions (2.16), (2.17), then we get formulas for approximate simulation of a potential bi-isotropic field on a plane. For a potential field of arbitrary dimension harmonic ξ is of the form

$$\xi(x) \doteq \frac{1}{\gamma}\begin{bmatrix} \lambda^{(1)} \\ \vdots \\ \lambda^{(k)} \end{bmatrix}(\eta\cos < x, \lambda > -\eta'\sin < x, \lambda >) =$$

$$= \frac{1}{\gamma}\begin{bmatrix} \lambda^{(1)} \\ \vdots \\ \lambda^{(k)} \end{bmatrix}\sqrt{-2\ln\alpha}\cos(< x, \lambda > +2\pi\beta), \qquad (2.18)$$

where $\gamma = \|\lambda\|$; η, η' are standard normal variables; α and β are independent and uniformly distributed in (0,1). Hence, actually the algorithm is reduced to approximate modelling of a scalar isotropic or, to be more precise, locally isotropic field in $\mathbb{R}^k$ and its coordinate-wise differentiation (cf. Chapter 1)

Analogue of harmonic (2.17) for simulation of solenoidal k-dimensional fields is of the form

$$\xi(x) = A(\lambda)\left\{\begin{bmatrix} \eta^{(1)} \\ \vdots \\ \eta^{(k)} \end{bmatrix}\cos < x, \lambda > + \begin{bmatrix} \eta'^{(1)} \\ \vdots \\ \eta'^{(k)} \end{bmatrix}\sin < x, \lambda >\right\} =$$

$$= A(\lambda)\begin{bmatrix} \sqrt{-2\ln\alpha^{(1)}}\cos\left(< x, \lambda > +2\pi\beta^{(1)}\right) \\ \vdots \\ \sqrt{-2\ln\alpha^{(k)}}\cos\left(< x, \lambda > +2\pi\beta^{(k)}\right) \end{bmatrix},$$

where $A(\lambda)A^T(\lambda) = E - L(\lambda)$; $\eta^{(i)}$, $\eta'^{(i)}$ are independent standard normal variables; $\alpha^{(i)}$ and $\beta^{(i)}$ are independent and uniformly distributed in (0,1). We can take $A(\lambda) = E - L(\lambda)$:

$$(E - L(\lambda))^T = E - L(\lambda), \quad (E - L(\lambda))^2 = E - L(\lambda).$$

because $L(\lambda)$ is the matrix of projection operator onto one-dimensional space which contains vector λ and $I - L(\lambda)$ is the projection matrix corresponding to orthogonal complement of the vector λ.

For three-dimensional solenoidal bi-isotropic fields we can also use the following formulas for simulation (see [53])

$$\begin{aligned}\xi(x) &= \left(\begin{bmatrix}\eta^{(1)}\\ \eta^{(2)}\\ \eta^{(3)}\end{bmatrix} \times \frac{\lambda}{\gamma}\right)\cos < x, \lambda > + \\ &+ \left(\begin{bmatrix}\eta'^{(1)}\\ \eta'^{(2)}\\ \eta'^{(3)}\end{bmatrix} \times \frac{\lambda}{\gamma}\right)\sin < x, \lambda >= \\ &= \begin{bmatrix}\sqrt{-2\ln\alpha^{(1)}}\cos\left(< x, \lambda > +2\pi\beta^{(1)}\right)\\ \sqrt{-2\ln\alpha^{(2)}}\cos\left(< x, \lambda > +2\pi\beta^{(2)}\right)\\ \sqrt{-2\ln\alpha^{(3)}}\cos\left(< x, \lambda > +2\pi\beta^{(3)}\right)\end{bmatrix} \times \frac{\lambda}{\gamma},\end{aligned}$$

where

$$a \times b = \left(\begin{vmatrix} a_2 & a_3 \\ b_2 & b_3 \end{vmatrix}, \begin{vmatrix} a_3 & a_1 \\ b_3 & b_1 \end{vmatrix}, \begin{vmatrix} a_1 & a_2 \\ b_1 & b_2 \end{vmatrix}\right)^T$$

is the vector product.

Remark 1. A peculiarity of the presented formulas for simulation is that the principle of spectrum partitioning is proposed to be used not only with respect to the absolute value $\gamma = \|\lambda\|$ (see, for example, [106], [107]), but with respect to the hemisphere of wave vector directions as well.

Remark 2. Formulas for simulation of partially homogeneous and locally homogeneous random fields can be found in [58]

Remark 3. Computer programs for simulation of certain classes of scalar and vector random fields are presented in paper [91].

Chapter 3

Convergence of spectral models of random fields in Monte Carlo methods

3.1. Introduction

Let us consider one of the principal issues encountered in solving various applied problems with use of numerical modelling of random processes and fields. Assume that a quantity $J = \mathbf{M}f(w)$ is estimated by Monte Carlo method, where f is a certain functional of realization of a random field w. By w_n we denote an approximate numerical model of the field w. For the justified use of the numerical model, i.e., for the use of the empirical mean

$$N^{-1} \sum_{i=1}^{N} f(w_n(i)),$$

as an estimate of the quantity J, where $w_n(i)$ are independent sample functions of w_n, the convergence of the expectations is required:

$$\mathbf{M}f(w_n) \to \mathbf{M}f(w) \quad \text{as } n \to \infty. \tag{3.1}$$

In general case, this convergence does not follow from convergence of w_n to w in the sense of convergence of finite-dimensional distributions or mean-square convergence. Obtaining efficient estimates for $|\mathbf{M}f(w_n) - \mathbf{M}f(w)|$ is often complicated. One way to prove convergence (3.1) for a wide class of functionals is to take advantage of the theory of weak convergence of probability measures.

Definition 1. Let sample functions of random fields $w, w_n, n = 1, 2, \ldots$, belong to a certain metric space S with probability one. *Weak convergence* of w_n to w in the space S is equivalent to convergence of expectations (3.1) for any functional f measurable on S such that random variables $f(w_n)$ are uniformly integrable and

$$\mathbf{P}(w \in D_f) = 0,$$

where D_f is a set of discontinuities of the functional f.

In this Chapter we study conditions of convergence (3.1) for different classes of functionals f and for different numerical models of random fields w_n. In particular, we consider weak convergence in spaces of continuously differentiable and integrable functions for spectral models of Gaussian fields. The results from papers [88], [92], [93], [98], are substantially used here.

Some well-known facts from the theory of weak convergence are presented below (cf., for example, [14], [13], [30], [54], [100]).

Let S be a complete separable metric space with a metric ρ, and $\mathcal{B}$ be the minimal σ-algebra containing all open sets. Consider the probability measures $\mathbf{P}$, $\mathbf{P}_n$, $n = 1, 2, \ldots$, on the measurable space $(S, \mathcal{B})$.

Definition 2. A sequence $\mathbf{P}_n$ is said to be *weakly convergent* to $\mathbf{P}$ (it will be denoted as $\mathbf{P}_n \Rightarrow \mathbf{P}$) if one of the following equivalent conditions is satisfied:

1) for any bounded real-valued continuous functional g on S we have

$$\int g \, d\mathbf{P}_n \to \int g \, d\mathbf{P};$$

2) for any continuous functional f on S, the probability measures $\mathbf{P}_n f^{-1}$ on the real line $\mathbb{R}$ converge in distribution to $\mathbf{P} f^{-1}$ (recall that convergence in distribution is equivalent to convergence of characteristic functions for each point and equivalent to convergence of distribution functions at any point, where the limit distribution function is continuous);

3) for any set $A \in \mathcal{B}$ such that $\mathbf{P}(\partial A) = 0$ we have $\mathbf{P}_n(A) \to \mathbf{P}(A)$;

4) the measures $\mathbf{P}_n$ converge to $\mathbf{P}$ in Prokhorov's metric:

$$L(\mathbf{P}_n, \mathbf{P}) \to 0,$$

where

$$L(\mathbf{Q}, \mathbf{P}) = \inf \{\varepsilon > 0 : \mathbf{P}(A) \le \mathbf{Q}(A^\varepsilon) + \varepsilon, \ \mathbf{Q}(A) \le \mathbf{P}(A^\varepsilon) + \varepsilon, A \in \mathcal{B}\},$$

$$A^\varepsilon = \{x \in S : \rho(x, A) < \varepsilon\}.$$

Definition 3. A sequence of random elements w_n is called weakly convergent to a random element w in the space S ($w_n \Rightarrow w$) if the corresponding measures on S are weakly convergent.

Convergence of expectations is of particular interest in Monte Carlo methods.

Theorem. *Let $w_n \Rightarrow w$ and f be a measurable functional such that $\mathbf{P}(w \in D_f) = 0$, where D_f is a set of discontinuities of the functional f. Then $f(w_n) \Rightarrow f(w)$ and under additional assumption of uniform integrability of random variables $f(w_n)$ we have*

$$\mathbf{M}f(w_n) \to \mathbf{M}f(w).$$

The Definition 1 of weak convergence is just a restatement of this theorem.

Definition 4. A family of probability measures Π is called *weakly compact* if a weakly convergent subsequence $\mathbf{P}_{n_k}$ can be extracted from an arbitrary sequence $\mathbf{P}_n \subset \Pi$: $\mathbf{P}_{n_k} \to \mathbf{P}$, where $\mathbf{P}$ is a probability measure. A family of random elements is called weakly compact if the corresponding family of probability measures is weakly compact.

Definition 5. A family of probability measures Π is said to be *dense* if for each $\varepsilon > 0$ there exists a compactum $K \subset S$ such that

$$\mathbf{P}(K) > 1 - \varepsilon, \quad \text{for all } \mathbf{P} \text{ from } \Pi.$$

Theorem (Yu.V. Prokhorov). *A family of probability measures in a separable complete metric space is weakly compact if and only if this family is dense.*

3.2. Conditions of weak convergence in spaces C and C^p

3.2.1. Weak convergence in the space of continuous functions

In this Item as a metric space S we will consider the space of continuous functions $C(X)$ defined on a compactum $X \subset \mathbb{R}^k$ with the norm $\|w\| = \sup\{|w(x)|,\ x \in X\}$.

Necessary and sufficient conditions for weak convergence in the space $C(X)$ were obtained by Yu. V. Prokhorov [100].

Theorem 3.1. *For a sequence of random fields $w_n(x)$ weakly converge to a field $w(x)$ in the space $C(X)$ it is necessary and sufficient that the finite-dimensional distributions of the fields w_n converge to the corresponding ones of the field w and the following condition be satisfied for all $a > 0$*

$$\sup_n \mathbf{P}\left\{\sup\left(|w_n(x) - w_n(x')|, |x - x'| < h\right) > a\right\} \to 0 \tag{3.2}$$

as $h \to 0$.

Remark. Condition (3.2) with

$$\sup_n \mathbf{P}\{|w_n(x_0)| > a\} \to 0 \quad \text{as } a \to \infty, \tag{3.3}$$

for some $x_0 \in X$, is equivalent to weak compactness of the sequence w_n in the space $C(X)$ (cf. [13]).

Theorem 3.1 will be used repeatedly hereinafter. In addition, in this Chapter we will refer to paper [52] more than once. It is appropriate to study this paper independently.

Sufficient conditions of weak compactness of a family of homogeneous random fields in terms of spectral moments were proved in [88].

Lemma 3.1. *Weak compactness in $C(X)$ of a sequence of homogeneous fields w_n with uniformly bounded variances ($\sup_n \mathbf{M}w_n^2 < \infty$) and spectral measures μ_n follows from condition*

$$\sup_n \int_{\mathbf{R}^k} |\lambda|^\beta \mu_n(d\lambda) < \infty, \tag{3.4}$$

where $\beta = 2([k/2]+1)$ ($[\ .\ ]$ denotes integer part of a number). For weak compactness of a sequence of Gaussian fields w_n it is sufficient that $\beta > 0$ for inequality (3.4).

Proof. Condition (3.3) is obviously fulfilled. So, to prove Lemma 1 we shall verify condition (3.2).

First, we consider the Gaussian case. Without loss of generality we may assume that $\beta \leq 2$. Then (3.4) implies

$$\sup_n \mathbf{M}|w_n(x+h) - w_n(x)|^2 \le c|h|^\beta.$$

and, clearly,

$$\sup_n \sup\{\mathbf{M}|w_n(x+h) - w_n(x)|^2, \quad |h| < h_0\} \le c h_0^\beta.$$

The latter inequality is sufficient for (3.2) in the Gaussian case (cf. [52]).

For non-Gaussian homogeneous fields the statement of Lemma 3.1 is a corollary of Theorem 2 from V.V. Yurinskiy's paper [126]. This theorem is presented below for $p = 2$ with allowance for the fact that mean square derivative at the same time is generalized stochastic derivative [59].

Theorem 3.2. *Let $w(x)$ be a measurable separable stochastically continuous random field defined on an open set $G \subset \mathbb{R}^k$ such that*

$$\mathbf{M}\int_G |w(x)|^2 dx < \infty.$$

Assume that there exist all mean square derivatives

$$\mathcal{D}_{m_1\ldots m_k} w(x) = \frac{\partial^m}{\partial x_1^{m_1} \ldots \partial x_k^{m_k}} w(x)$$

of order $m = m_1 + \ldots + m_k = [k/2] + 1$ and

$$\mathbf{M}\int_G |\mathcal{D}_{m_1\ldots m_k} w(x)|^2 dx < \infty.$$

Then sample functions of the random field $w(x)$ are continuous with probability one. Moreover, if G is an open cube with edge a, we have the inequality

$$|w(x) - w(x')| \le c|x - x'|_*^{1/2} \|w\|_G, \tag{3.5}$$

which is satisfied with probability one for all x, x' such that $|x - x'|_* < a/2$. Here

$$\|w\|_G^2 = \int_G \Big(|w(x)|^2 + \sum_{m_1+\ldots+m_k=m} |\mathcal{D}_{m_1\ldots m_k} w(x)|^2 \Big) dx < \infty,$$

$$|x|_* = \max\{|x_i|, i = 1, \ldots, k\},$$

and constant c depends on the size of the cube and dimension k.

To prove weak compactness of the sequence w_n in Lemma 3.1 it is sufficient to establish that

$$\sup_n \mathbf{P}\left\{\sup_{x,x'\in G}(|w_n(x)-w_n(x')|,\ |x-x'|_* < h) > a\right\} \underset{h\to 0}{\to} 0,$$

where $a > 0$ and G is an open cube containing X. Inequality (3.5) implies that if h is small enough, the following is valid

$$\mathbf{P}\left\{\sup_{x,x'\in G}(|w_n(x)-w_n(x')|,\ |x-x'|_* < h) > a\right\} \le$$
$$\le \mathbf{P}\{c^2 h\|w_n\|_G^2 > a^2\} \le c^2 h\mathbf{M}\|w_n\|_G^2/a^2.$$

Condition (3.4) ensures existence of the corresponding mean square derivatives and enables us to write

$$\sup_n \mathbf{M}\|w_n\|_G^2 = \sup_n \int_G \Big(\mathbf{M}|w_n(x)|^2 +$$
$$+ \sum_{m_1+\ldots+m_k=m} \mathbf{M}|\mathcal{D}_{m_1\ldots m_k} w_n(x)|^2\Big)dx \le A < \infty.$$

(Insertion of mathematical expectation under the integral sign is valid in accordance with Theorem 2 from Section 3.3 of [30].) Hence, we have

$$\sup_n \mathbf{P}\left\{\sup_{x,x'\in X}(|w_n(x)-w_n(x')|, |x-x'|_* < h) > a\right\} \le c^2 Ah/a^2$$

for sufficiently small h, and the weak convergence condition (3.2) is fulfilled.

Lemma 3.1 has been proved.

3.2.2. Weak convergence of probability measures in spaces of continuously differentiable functions

In this Item a criterion for weak convergence of probability measures in spaces of continuously differentiable functions $C^p(X)$, $X \subset \mathbb{R}^k$, is presented (cf. [93], [98], [122]). As a corollary, "spectral" conditions of weak compactness of a sequence of homogeneous random fields and central limit theorem in $C^p(X)$ are obtained.

Let X be a bounded closed domain in $\mathbb{R}^k$, $C^p(X)$ be a space of functions with continuous partial derivatives of an order lower or equal to p, $p \in \{0, 1, 2, \dots\}$, and with norm

$$\|w\| = \max_{|m| \le p} \sup_{x \in X} |\mathcal{D}^m w(x)|,$$

where

$$\mathcal{D}^m w(x) = \frac{\partial^{|m|}}{\partial x_1^{m_1} \dots \partial x_k^{m_k}} w(x),$$
$$m = (m_1, \dots, m_k), \quad |m| = m_1 + \dots + m_k.$$

We consider measurable space $C^p(X)$ with σ-algebra generated by open sets. By $\mathbf{P}_n$ we denote a sequence of probability measures in the space $C^p(X)$.

Lemma 3.2. *A sequence of the probability measures $\mathbf{P}_n$ is weakly compact in $C^p(X)$, if and only if the following conditions are satisfied:*

1) *for a certain $x_0 \in X$ and any $\varepsilon > 0$ there exists M such that*

$$\max_{|m| \le p} \sup_n \mathbf{P}_n\{|\mathcal{D}^m w(x_0)| > M\} \le \varepsilon; \tag{3.6}$$

2) *for any $a > 0$ and multiindex m such that $|m| = p$,*

$$\lim_{h \to 0} \sup_n \mathbf{P}_n\{\Delta_w^m(h) \ge a\} = 0, \tag{3.7}$$

where

$$\Delta_w^m(h) = \sup_{|x-y|<h} |\mathcal{D}^m[w(x) - w(y)]|.$$

The proof of Lemma 3.2 is based on the following statement.

Lemma 3.3. *For a set $A \subset C^p(X)$ be relatively compact (i.e., its closure is a compactum), it is necessary and sufficient that*

1) *for a certain $x_0 \in X$*

$$\max_{|m| \le p} \sup_{w \in A} |\mathcal{D}^m w(x_0)| < \infty; \tag{3.8}$$

2) *for all m such that $|m| = p$,*

$$\lim_{h\to 0} \sup_{w\in A} \Delta_w^m(h) = 0. \tag{3.9}$$

Remark. In the formulations of Lemmas 3.2 and 3.3 point x_0 can be varied for different derivatives: $x_0 = x_0(m)$.

A simple corollary of Lemma 3.2 is the following theorem.

Theorem 3.3. *Convergence of finite-dimensional distributions and conditions (3.6) and (3.7) are necessary and sufficient for weak convergence of probability measures $\mathbf{P}_n$ to a probability measure $\mathbf{P}$ in $C^p(X)$.*

Proof of Lemma 3.3. Necessity of conditions (3.8), (3.9) follows from existence of a finite ε-net for arbitrary $\varepsilon > 0$.

Proof of sufficiency will be divided into two steps.

Step 1. Consider an arbitrary sequence of continuously differentiable functions $w_i \in C^p(X)$. Let us show that for any $h > 0$, there exists $x^* \in X$ and a subsequence w_i^* of the sequence w_i such that

$$|x_{ij}^* - x^*| \leq h,$$

where

$$x_{ij}^* = \arg \|w_i^* - w_j^*\|.$$

By $\arg \|w\|$ we will denote a point $x \in X$ (arbitrarily chosen in the case of ambiguity) for which the following holds

$$\|w\| = \max_{|m|\leq p} |\mathcal{D}^m w(x)|.$$

Denote

$$x_{ij} = \arg \|w_i - w_j\|.$$

As X is a compactum, then for the sequence x_{1i} there exists a limit point $x_1 \in X$. We choose a subsequence $w_i^{(1)}$ of the sequence w_i, such that

$$w_1^{(1)} = w_1, \quad |x_{1j}^{(1)} - x_1| \le h/2,$$

where

$$x_{ij}^{(1)} = \arg \|w_i^{(1)} - w_j^{(1)}\|.$$

Let $x_2 \in X$ be a limit point of the sequence $x_{2j}^{(1)}$. We choose next a subsequence $w_i^{(2)}$ of the sequence $w_i^{(1)}$, such that

$$w_1^{(2)} = w_1^{(1)} = w_1, \quad w_2^{(2)} = w_2^{(1)},$$
$$|x_{2j}^{(2)} - x_2| \le h/2, \quad j > 2,$$

where

$$x_{ij}^{(2)} = \arg \|w_i^{(2)} - w_j^{(2)}\|.$$

Continuing this procedure, we will obtain a series of sequences

$$\{w_i^{(m)}\} \subset \{w_i^{(m-1)}\},$$

where

$$w_i^{(m)} = w_i^{(m-1)}, \quad i = 1, \dots, m,$$
$$|x_{mj}^{(m)} - x_m| \le h/2, \quad j > m,$$
$$x_{ij}^{(m)} = \arg \|w_i^{(m)} - w_j^{(m)}\|.$$

Consider "diagonal" sequence $\tilde{w}_i = w_i^{(i)}$. Evidently, this sequence obeys the inequality

$$|\tilde{x}_{mj} - x_m| \le h/2, \quad j > m,$$

where

$$\tilde{x}_{ij} = \arg \|\tilde{w}_i - \tilde{w}_j\|.$$

Let $x^* \in X$ be a limit point of the sequence x_i and $x_i^* = x_{k_i}$ be a subsequence such that

$$|x_i^* - x^*| \le h/2.$$

Then the sequence $w_i^* = \tilde{w}_{k_i}$ possesses the required property:

$$|x_{ij}^* - x^*| \le |x_{ij}^* - x_m^*| + |x_m^* - x^*| \le h, \quad m = \min(i, j).$$

Step 2. Assume that the set A is not relatively compact. This implies (see, for example, [51]), that there exists $a > 0$ and a sequence $w_i \in A$ such that

$$\|w_i - w_j\| > a \quad \text{for } i \neq j. \tag{3.10}$$

To prove Lemma 3.3 means to show that conditions (3.8), (3.9) and (3.10) are incompatible.

It follows from conditions (3.8), (3.9) that the quantity

$$\sup_{x \in X} \sup_{w \in A} \max_{|m| \leq p} |\mathcal{D}^m w(x)| \tag{3.11}$$

is bounded (see Appendix 2 in [13]).

On the other hand, from conditions (3.8), (3.9) it follows that there exists $h > 0$ such that

$$\sup_{|x-y|<h} \sup_{w \in A} \max_{|m| \leq p} |\mathcal{D}^m (w(x) - w(y))| < a/3.$$

According to Step 1, for the given value of h, we choose $x^* \in X$ and a subsequence $w_i^* \in A$ of the sequence w_i such that

$$|x_{ij}^* - x^*| \leq h,$$

where

$$x_{ij}^* = \arg \|w_i^* - w_j^*\|.$$

Inequality (3.10) implies

$$a < \|w_i^* - w_j^*\| = \max_{|m| \leq p} \left|\mathcal{D}^m \left(w_i^*(x_{ij}^*) - w_j^*(x_{ij}^*)\right)\right| =$$

$$= \max_{|m| \leq p} \Big|\mathcal{D}^m \Big[\left(w_i^*(x_{ij}^*) - w_i^*(x^*)\right) + \left(w_i^*(x^*) - w_j^*(x^*)\right) +$$

$$+ \left(w_j^*(x^*) - w_j^*(x_{ij}^*)\right)\Big]\Big|$$

for $i \neq j$ and, hence,

$$\max_{|m| \leq p} \left|\mathcal{D}^m \left[w_i^*(x^*) - w_j^*(x^*)\right]\right| \geq a/3.$$

Thus, the quantity

$$\sup_i \max_{|m| \leq p} |\mathcal{D}^m w_i^*(x^*)|$$

is unbounded, which is in contradiction with the boundedness of (3.11). Lemma 3.3 has been proved.

Proof of Lemma 3.2. Necessity. According to Prokhorov's theorem (cf. [100] and Section 3.1), weak compactness of the sequence of measures $\mathbf{P}_n$ means that for any $\varepsilon > 0$ there exists a compactum $K \subset C^p(X)$ such that

$$\sup_n \mathbf{P}_n\{C^p(X) - K\} \le \varepsilon. \tag{3.12}$$

Let us fix some $a > 0$, $\varepsilon > 0$ and choose a compactum K so that (3.9) is satisfied. According to Lemma 3.3, there exists $x_0 \in X$ such that

$$K \subset \bigcap_{|m| \le p} \{w : \mathcal{D}^m |w(x_0)| \le M\}$$

for some M and

$$K \subset \bigcap_{|m|=p} \{w : \Delta_w^m(h) < a\}$$

with sufficiently small h. Hence (3.12) implies (3.6), (3.7).

Sufficiency. Assume that conditions (3.6), (3.7) are valid. Let us consider the sets

$$A_m = \{w : |D^m w(x_0)| \le M\}, \quad |m| \le p,$$

where M is chosen so that

$$\sum_{|m| \le p} \sup_n \mathbf{P}\{w_n \notin A_m\} \le \varepsilon/2.$$

Moreover, let us consider the sets

$$B_m^s = \{w : \Delta_w^m(h_s) < 1/s\}, \quad |m| = p, \quad s = 2, 3, \ldots,$$

where h_s are chosen so that

$$\sum_{|m|=p} \sup_n \mathbf{P}\{w_n \notin B_m^s\} \le \varepsilon/2^s.$$

The closure of the set $\left(\bigcap_{|m| \le p} A_m\right) \cap \left(\bigcap_{s=2}^{\infty} \bigcap_{|m|=p} B_m^s\right)$ will be denoted by K. According to Lemma 3.3, the set K is a compactum and, evidently, the following is valid

$$\sup_n \mathbf{P}_n\{C^p(X) - K\} \leq \varepsilon.$$

Lemma 3.2 has been proved.

We now present two applications of the statements above to demonstrate a scope of studying weak convergence in $C^p(X)$ by adaptation of corresponding results for the space $C(X)$.

Lemma 3.4 (Sufficient conditions for weak compactness of a sequence of homogeneous fields in $C^p(X)$). *Let X be a compactum in $\mathbb{R}^k$. A sequence of homogeneous fields $w_n(x)$, $x \in X$, with spectral measures μ_n is weakly compact in $C^p(X)$ (i.e., the corresponding sequence of probability measures is weakly compact), if*

$$\sup_n \mu_n(\mathbb{R}^k) < \infty,$$

$$\sup_n \int_{\mathbb{R}^k} |\lambda|^{2p+\beta} \mu_n(d\lambda) < \infty, \tag{3.13}$$

where $\beta = 2([k/2]+1)$ and $[\ .\ ]$ denotes integer part of a number. If, moreover, the fields w_n are Gaussian, then $\beta > 0$ in (3.13) is sufficient for weak compactness in $C^p(X)$.

Proof. For $p = 0$ the Lemma has been already proved (see Lemma 3.1). From (3.13) it follows that the random fields $\mathcal{D}^m w(x)$, $|m| = p$, obtained by differentiation of sample functions, are weakly compact in $C(X)$, which, in turn, means that condition 2 of Lemma 3.2 is satisfied. Moreover, condition 1 of Lemma 3.2 is satisfied, too. Thus, the sequence of the fields $w_n(x)$ is weakly compact in $C^p(X)$.

Remark. For stationary processes the following condition

$$\sup_n \int_{\mathbb{R}} |\lambda|^{1+\alpha} \mu_n(d\lambda) < \infty, \quad \alpha > 0,$$

is sufficient for weak compactness of the sequence $w_n(x)$ in $C(X)$, $x \in X \subset \mathbb{R}$ (it follows from the Kolmogorov criterion for continuity of random processes). Therefore, the condition

$$\sup_n \int_{\mathbb{R}} |\lambda|^{2p+\beta} \mu_n(d\lambda) < \infty, \quad \beta > 1,$$

is sufficient for weak compactness of the processes w_n in $C^p(X)$, $X \subset \mathbb{R}$.

Task 3.1. Give an answer to the question: can we weaken the condition $\beta = 2([k/2]+1)$ in Lemma 3.4 and require only $\beta > k$?

Lemma 3.5 (Central limit theorem in the space $C^p(X)$). *Let X be a bounded closed domain in $\mathbb{R}^k$; ε be a random element with values in the space $C^p(X)$, $\mathbf{M}\varepsilon(x) = 0$, $x \in X$; ε_i be independent realizations of ε and*

$$w_n(x) = n^{-1/2} \sum_{i=1}^{n} \varepsilon_i(x).$$

The sequence w_n weakly converge to a Gaussian random element in $C^p(X)$ if the following conditions are fulfilled:

a) $\max\limits_{|m| \le p} \sup\limits_{x \in X} \mathbf{M}[\mathcal{D}^m \varepsilon(x)]^2 < \infty$;

b) *for every multiindex $m = (m_1, \dots, m_k)$ such that $|m| = p$, there exist a random variable $G(\omega) > 0$, $\mathbf{M}G^2 < \infty$, and a metric ρ on X, continuous with respect to the Euclidean metric, so that*

$$|\mathcal{D}^m[\varepsilon(x,\omega) - \varepsilon(y,\omega)]| < G(\omega)\rho(x,y), \quad x, y \in X,$$

$$\int_0^1 (\ln N_\rho(\delta))^{1/2} d\delta < \infty.$$

Here $N_\rho(\delta)$ denotes minimal number of balls with radius smaller or equal to δ in the metric ρ, which cover the set X.

Proof. In view of the Jain–Marcus theorem [45], conditions a) and b) of the Lemma imply weak convergence in $C(X)$ of random elements (obtained by differentiation of trajectories) $\mathcal{D}^m w_n$, $|m| = p$, to corresponding Gaussian elements. It means that condition 2 of Lemma 3.2 is satisfied. Evidently, condition 1 of Lemma 3.2 is also fulfilled. Hence, according to Lemma 3.2 the sequence w_n is weakly compact in $C^p(X)$ and, therefore, weakly converges to a Gaussian element in $C^p(X)$.

The central limit theorem in spaces of continuously differentiable functions finds its use in Monte Carlo methods, in particular, in studying convergence of numerical models of random functions (see Section 4.3) and for justification of the method of dependent tests (see, for example, [97], [121] and Chapter 4).

3.3. Convergence of spectral models of Gaussian homogeneous fields

Due to universal character and simplicity of simulation algorithms, spectral models of Gaussian fields find use in solving a wide class of applied problems. Sufficient conditions of weak convergence of spectral models which will be described in this Section enable us to give an advice on choosing parameters of algorithms, and provide asymptotic unbiasedness of corresponding estimates obtained by Monte Carlo method (see formula (3.1)).

First we recall basic concepts concerning spectral models of random fields.

3.3.1. Spectral models

Let $w(x)$, $x \in \mathbb{R}^k$, be a homogeneous Gaussian field with zero mean, unit variance and spectral measure μ. For numerical simulation of the field w we consider approximate models of the form

$$w_n(x) = \sum_{j=1}^{m} a_j \left[\xi_j \cos < \lambda_j, x > + \eta_j \sin < \lambda_j, x >\right], \tag{3.14}$$

which will be referred to as *spectral.* Here $a_j > 0$; (ξ_j, η_j) are independent, identically distributed random vectors such that $\mathbf{M}\xi_j = \mathbf{M}\eta_j = \mathbf{M}\xi_j\eta_j = 0$, $\mathbf{D}\xi_j = \mathbf{D}\eta_j = 1$; λ_j are vectors of the dimension k; $< \cdot, \cdot >$ denotes the scalar product in $\mathbb{R}^k$. The quantities m, a_j, λ_j in formula (3.14) and hereafter depend on n, however, for brevity we will omit the index n.

Vectors λ_j in formula (3.14) can be random (*randomized models*) as well as non-random (*non-randomized models*). Randomized models enable to reproduce exactly correlation function and spectral measure μ of the simulated field w.

Evidently, the fields w_n in (3.14) are homogeneous. Strict homogeneity (the invariance of finite-dimensional distributions with respect to shifts) is equivalent to isotropy of the vectors (ξ_j, η_j). In this case, representation

$$w_n(x) = \sum_{j=1}^{m} a_j r_j \cos(< \lambda_j, x > + \varphi_j),$$

where $r_j^2 = \xi_j^2 + \eta_j^2$ and φ_j are random variables uniformly distributed over the segment $(0, 2\pi)$ and independent of r_j, can prove to be more efficient. Henceforth, unless otherwise stated, we will assume that the vectors (ξ_j, η_j) are Gaussian, which ensures that the fields w_n are Gaussian, provided λ_j are non-random, and that one-dimensional distribution is Gaussian for randomized models. In this case, for modelling of r_j, we can use the expression

$$r_j = (-2 \ln \alpha_j)^{1/2},$$

where α_j are random variables uniformly distributed on $(0, 1)$.

Let

$$\Lambda = \mathbb{R}^k = \sum_{j=1}^{m} \Lambda_j, \quad \Lambda_j \cap \Lambda_i = \emptyset \text{ for } j \neq i$$

be a partitioning (dependent on n) of the spectral space Λ. We consider the following spectral models.

Model A: $a_j^2 = \mu(\Lambda_j)$, $\lambda_j \in \Lambda_j$, λ_j are non-random.

Model B: $a_j^2 = \mu(\Lambda_j)$ and λ_j are random vectors with distributions concentrated in the corresponding domains Λ_j and induced by the spectral measure μ.

Model C: $a_j^2 = 1/m$ and λ_j are independent random vectors distributed according to the spectral measure μ.

Models B, C are randomized. The fields w_n, corresponding to these models have the correlation function which coincides with that of the field w being simulated.

Spectral decomposition of the homogeneous field has the form

$$w(x) = \int_\Lambda \cos < \lambda, x > \xi(d\lambda) + \int_\Lambda \sin < \lambda, x > \eta(d\lambda),$$

where ξ and η are orthogonal stochastic measures such that

$$\mathbf{M}\xi^2(d\lambda) = \mathbf{M}\eta^2(d\lambda) = \mu(d\lambda), \quad \mathbf{M}\xi(d\lambda)\eta(d\lambda) = 0.$$

Thus, models A–C are characterized by different approximations of the stochastic measures.

For Model C, convergence of finite-dimensional distributions to the Gaussian distributions of the limit field w follows from the central limit theorem. Conditions

$$\max_{j\leq m}\mu(\Lambda_j)\to 0,\quad \text{as } n\to\infty, \tag{3.15}$$

are sufficient for convergence of finite-dimensional distributions for Model B, while for Model A the following conditions are sufficient

$$\begin{aligned}&\Lambda_m\subset\{|\lambda|\geq c_n\},\quad c_n\to\infty,\\ &\max_{j<m}\operatorname{diam}(\Lambda_j)\to 0\quad \text{as } n\to\infty.\end{aligned} \tag{3.16}$$

Exercise 3.1. Prove yourself just formulated sufficient conditions of convergence of finite-dimensional distributions for spectral Models A and B.

Remarks. 1. Condition (3.15) is not sufficient for convergence of finite-dimensional distributions for Model A.

2. If the spectral measure μ does not contain a discrete component, then condition (3.16) implies (3.15) and condition (3.15) is not only sufficient, but also necessary for convergence of finite-dimensional distributions in Model B.

Mean square convergence of spectral models has been studied in [72], [120], [56].

3.3.2. Weak convergence of spectral models in spaces of continuously differentiable functions

The following statement is a simple corollary of Lemma 3.4.

Lemma 3.6. *Let X be a compactum in $\mathbb{R}^k$. Weak convergence in $C^p(X)$ of spectral model (3.14) to a homogeneous Gaussian field w follows from the condition*

$$\int_\Lambda |\lambda|^{\beta+2p}\mu(d\lambda)<\infty, \tag{3.17}$$

where

a) $\beta > 0$ *for* Model A *subject to the condition (3.16) and additional restrictions on the increase of* λ_m, *e.g.,* $|\lambda_m| = c_n$;

b) $\beta = 2[k/2] + 2$ *for* Model B *under the condition (3.15) and for* Model C *([.] denotes integer part of a number).*

Taking advantage of peculiar features of randomized models, we can expand sufficient conditions for weak convergence.

Theorem 3.4. *If condition (3.17) is satisfied for some* $\beta > 0$, *then the randomized spectral* Model B, *satisfying conditions (3.15) and (3.16), weakly converges in* $C^p(X)$ *to a homogeneous Gaussian field* w *with the spectral measure* μ.

Proof. Weak convergence in $C^p(X)$ is equivalent to convergence of finite-dimensional distributions and weak compactness. Convergence of the finite-dimensional distributions of the fields w_n to the Gaussian distributions follows from condition (3.15). Let us verify weak compactness of w_n.

First, we consider the case $p = 0$. In this case weak compactness is equivalent to convergence

$$\sup_n \mathbf{P}\left\{ \sup_{x,y:|x-y|<h} |w_n(x) - w_n(y)| > a \right\} \underset{h\to 0}{\to} 0 \tag{3.18}$$

for any $a > 0$. To establish (3.18), we shall take advantage of the following fact.

Proposition 1. *If two sequences of random fields* $v_n(x)$ *and* $u_n(x)$ *are weakly compact in* $C(X)$ *and mutually independent, then the sequence* $w_n(x) = v_n(x) + u_n(x)$ *is weakly compact, too.*

Proof of Proposition 1. Let us denote

$$f = \sup\{|v_n(x) - v_n(y)|, \quad |x-y| < h\},$$
$$g = \sup\{|u_n(x) - u_n(y)|, \quad |x-y| < h\}.$$

Proposition 1 follows from the equations below:

$$\begin{aligned}
&\mathbf{P}\{\sup(|w_n(x) - w_n(y)|, \quad |x-y| < h) > a\} \leq \\
&\leq \mathbf{P}\{f + g > a\} = 1 - \mathbf{P}\{f + g \leq a\} \leq \\
&\leq 1 - \mathbf{P}\{f \leq a/2, g \leq a/2\} = 1 - \mathbf{P}\{f \leq a/2\}\mathbf{P}\{g \leq a/2\} = \\
&= 1 - (1 - \mathbf{P}\{f > a/2\})(1 - \mathbf{P}\{g > a/2\}) = \\
&= \mathbf{P}\{f > a/2\} + \mathbf{P}\{g > a/2\} - \mathbf{P}\{f > a/2\}\mathbf{P}\{g > a/2\} \leq \\
&\leq \mathbf{P}\{f > a/2\} + \mathbf{P}\{g > a/2\}.
\end{aligned}$$

Thus, for the proof of Theorem 3.4, it will suffice to prove weak compactness of the sequences $v_n(x)$ and $u_n(x)$:

$$u_n(x) = \sum_{j=1}^{m-1} a_j \left[\xi_j \cos < \lambda_j, x > + \eta_j \sin < \lambda_j, x >\right],$$

$$v_n(x) = a_m \left[\xi_m \cos < \lambda_m, x > + \eta_m \sin < \lambda_m, x >\right].$$

Weak compactness of v_n follows from Proposition 2.

Proposition 2. *The sequence of random fields*

$$a_m r \cos\left(< \lambda_m, x > + \varphi_m\right),$$

where $a_m \to 0$ as $m \to \infty$ and r, λ_m, φ_m are arbitrary random variables, is weakly compact (and, hence, weakly converges to zero).

Proof of Proposition 2. It is required to show that

$$\sup_m \mathbf{P}\{\Delta_m(h) > a\} \to 0 \quad \text{as } h \to 0,$$

where

$$\Delta_m(h) = a_m r \sup_{|x-y|<h} \left|\cos(< \lambda_m, x > + \varphi_m) - \cos(< \lambda_m, y > + \varphi_m)\right|.$$

Let us fix any $\varepsilon > 0$. Since

$$\Delta_m(h) \leq 2 a_m r \quad \text{and } a_m \to 0,$$

there exists m_* such that

$$\sup_{m>m_*} \mathbf{P}\{\Delta_m(h) > a\} \le \varepsilon \quad \text{for all } h. \tag{3.19}$$

We choose $C > 0$ such that

$$\sup_{m\le m_*} \mathbf{P}\{r|\lambda_m| > C\} \le \varepsilon, \tag{3.20}$$

and then we take $h_0 > 0$ such that the inequality

$$\sup_{m\le m_*} a_m C h_0 < a. \tag{3.21}$$

is satisfied. To proof the proposition, it is enough to show that for all $h < h_0$

$$\sup_m \mathbf{P}\{\Delta_m(h) > a\} \le \varepsilon.$$

In view of (3.19) we have

$$\sup_m \mathbf{P}\{\Delta_m(h) > a\} \le \max\Big(\varepsilon, \sup_{m\le m_*} \mathbf{P}\{\Delta_m(h) > a\}\Big).$$

Taking into account inequalities (3.20), (3.21) and

$$\Delta_m(h) \le a_m r|\lambda_m|h,$$

we get for $h < h_0$

$$\begin{aligned}\sup_{m\le m_*} \mathbf{P}\{\Delta_m(h) > a\} &= \sup_{m\le m_*} \Big[\mathbf{P}\left\{\Delta_m(h) > a, r|\lambda_m| > C\right\} + \\ &\quad + \mathbf{P}\left\{\Delta_m(h) > a, r|\lambda_m| \le C\right\}\Big] \le \varepsilon + 0.\end{aligned}$$

This completes the proof of Proposition 2.

Now we return to the proof of Theorem 3.4. Let us prove weak convergence of u_n. Consider a random field

$$\tilde{u}_n(x) = \sum_{j=1}^{m-1} a_j \left[\xi_j \cos < \lambda_j, x > + \eta_j \sin < \lambda_j, x >\right],$$

where an arbitrary set $\lambda_j \in \Lambda_j$, $j = 1, \ldots, n-1$, is fixed. Since for any $\beta \in (0, 2]$, there exists $c(\beta)$ such that $1 - \cos\alpha \le c(\beta)|\alpha|^\beta$, then for $\beta \le 2$ we have

$$\mathbf{M}\left|\tilde{u}_n(x+\vec{h}) - \tilde{u}_n(x)\right|^2 = 2\left[\tilde{R}_n(0) - \tilde{R}_n(\vec{h})\right] =$$
$$= 2\int_\Lambda \left(1 - \cos < \lambda, \vec{h} >\right) \tilde{\mu}_n(d\lambda) \leq 2c(\beta) \int_\Lambda | < \lambda, \vec{h} > |^\beta \tilde{\mu}_n(d\lambda) \leq$$
$$\leq 2c(\beta)|\vec{h}|^\beta \int_\Lambda |\lambda|^\beta \tilde{\mu}_n(d\lambda).$$

Here $\tilde{R}_n$ is correlation function, and $\tilde{\mu}_n$ is spectral measure of the field $\tilde{u}_n(x)$ concentrated at the points λ_j with weights $\mu(\Lambda_j)$. From condition (3.16) it follows that

$$\int_\Lambda |\lambda|^\beta \tilde{\mu}_n(d\lambda) \to \int_\Lambda |\lambda|^\beta \mu_n(d\lambda) \quad \text{as } n \to \infty,$$

and, hence,

$$\sup\left\{\mathbf{M}|\tilde{u}_n(x+\vec{h}) - \tilde{u}_n(x)|^2, \ |\vec{h}| < h\right\} \leq Ch^\beta, \quad C > 0. \tag{3.22}$$

For fixed $\{\lambda_j\}$ the fields $\tilde{u}_n(x)$ are Gaussian and in view of the results from paper [52], relation (3.22) implies

$$\mathbf{P}\left\{\sup_{x,y:|x-y|<h} |\tilde{u}_n(x) - \tilde{u}_n(y)| > a\right\} \leq \varphi(h, a), \tag{3.23}$$

where $\varphi(h, a) \to 0$ as $h \to 0$. Since the right-hand sides of (3.22) and (3.23) are independent of n, we have

$$\sup_n \mathbf{P}\left\{\sup_{x,y:|x-y|<h} |\tilde{u}_n(x) - \tilde{u}_n(y)| > a\right\} \leq \varphi(h, a). \tag{3.24}$$

It remains to note that, in view of an arbitrary choice of λ_j in (3.23), this inequality is also valid for a random choice of λ_j. Weak compactness of the sequence $u_n(x)$ in $C(X)$ and the theorem for $p = 0$ have been proved.

For $p > 0$ we present only a sketch of the proof. To verify weak compactness of w_n in $C^p(X)$, we use Lemma 3.2. Conditions (3.6) are easily checked, and the proof of validity of (3.7) practically repeats the proof for the case $p = 0$.

Remarks. 1. As it is easy to see from the proof, Theorem 3.4 remains valid, if, instead of condition (3.16), we impose the requirement that the number of unbounded domains in partitioning of the spectral space be uniformly (with respect to n) bounded, while for bounded domains, the following be valid

$$\max_{\operatorname{diam}(\Lambda_j)<\infty} \operatorname{diam}(\Lambda_j) \to 0 \quad \text{as } n \to \infty. \tag{3.25}$$

Moreover, condition (3.16) may be replaced by (3.15) with additional requirement

$$\sup_n \sup_{\lambda_j \in \Lambda_j} \sum_{\operatorname{diam}(\Lambda_j)<\infty} |\lambda_j|^{\beta+2p} \mu(\Lambda_j) < \infty. \tag{3.26}$$

In particular, partitioning of the spectral space into "rings":

$$\Lambda_j = \{|\lambda| \in [d_j, d_{j+1})\}, \quad j = 1, \ldots, m-1,$$
$$\Lambda_m = \{|\lambda| \geq d_m\}, \quad 0 = d_1 < d_2 < \ldots < d_m < \infty,$$

under conditions

$$\max_{j=2,\ldots,m} |d_j - d_{j-1}| \to 0, \qquad \int_\Lambda |\lambda|^{\beta+2p} \mu(d\lambda) < \infty$$

satisfies (3.26), while condition (3.25) is not fulfilled.

2. Condition (3.17) for $\beta > 0$ is sufficient for weak convergence of spectral Model C in the space $C^p(X)$. It follows from the central limit theorem in $C^p(X)$ (see Lemma 3.5) and inequality

$$|\sin < \lambda, x > | \leq |\lambda|^\beta |x|^\beta \frac{\ln^\theta(e + |\lambda|)}{|\ln |x||^\theta},$$

where $|x| < 1$, $\beta \in [0,1]$, $\theta \in [0,1]$ (see also [56]).

3. Some other results on convergence of spectral models of random fields (including vector-valued fields) can be found in [16], [56], [57].

Task 3.2. Prove or refute the following hypothesis:

Hypothesis. As sample functions of the fields w_n are smooth, we may assume that convergence of finite-dimensional distributions and corresponding continuous differentiability of the limit field w,

$$\mathbf{P}\{w \in C^p(X)\} = 1.$$

will suffice for weak convergence of randomized spectral models (3.14) in $C^p(X)$.

3.3.3. On weak convergence in spaces of integrable functions

A.V. Ivanov's criterion [44] of weak convergence of random functions in $L^p(X)$, $X \subset \mathbb{R}^k$, enables us to verify weak convergence of spectral models in spaces L^p of integrable functions. In particular, it can be easily shown that convergence of finite-dimensional distributions of $w_n(x)$ to the Gaussian ones, and the fact that random variables ξ_j, η_j are Gaussian in (3.14), ensure weak convergence of spectral model (3.14) in $L^p(X)$, where X is a compactum in $\mathbb{R}^k$, $p \geq 1$.

Below we present the mentioned result obtained by A.V. Ivanov.

Let $X = [0,1]^k$ and $L^p(X)$ be a space of real measurable functions f on X, such that

$$\|f\| = \int\limits_X |f(x)|^p dx < \infty.$$

Theorem 3.5. *Let $\xi(x)$ and $\xi_n(x)$ be random fields ($x \in X$, $n = 1, 2, \dots$) with sample functions in $L^p(X)$, $p \geq 1$. If the following conditions are fulfilled:*

1) *finite-dimensional distributions of ξ_n converge to corresponding finite-dimensional distributions of ξ,*

2) $\mathbf{M}|\xi_n(x)|^p \to \mathbf{M}|\xi(x)|^p$ *for all $x \in X$,*

3) $\sup\limits_n \sup\limits_{x \in X} \mathbf{M}|\xi_n(x)|^p < \infty$,

then the fields ξ_n weakly converge to ξ in $L^p(X)$.

3.4. Convergence of conditional spectral models

Let us consider a spectral model

$$w_N(x) = \sum_{n=1}^{N} c_n \left[\xi_n \cos < \lambda_n, x > + \xi'_n \sin < \lambda_n, x >\right], \tag{3.27}$$

$$c_n > 0, \quad \sum_{n=1}^{N} c_n^2 < \infty,$$

of a Gaussian homogeneous random field $w(x)$, $x \in \mathbb{R}^k$, with mean zero, correlation function $R(x) = \mathbf{M}w(y)w(x+y)$, and spectral measure $\mu(d\lambda)$,

$$R(x) = \int \cos < \lambda, x > \mu(d\lambda).$$

Here ξ_n, ξ'_n are independent standard normal variables, while vectors $\lambda_n \in \mathbb{R}^k$ are random for randomized model or deterministic for non-randomized model.

We denote by $w^b(x)$ the Gaussian random field $w(x)$ provided that it hits the points

$$w(x_j) = b_j, \quad j \in \{1, 2, \ldots, J\}.$$

Hereafter we assume that correlation matrix of the random vector $(w(x_1), \ldots, w(x_J))$ is a nonsingular one.

Spectral model (3.27), where the vector

$$\zeta = (\xi_1, \ldots, \xi_N, \xi'_1, \ldots, \xi'_N)^T$$

is a Gaussian one with expectation A^+b and correlation matrix $I - A^+A$,

$$A = \begin{bmatrix} d_{11} & \ldots & d_{1N} & d'_{11} & \ldots & d'_{1N} \\ & & & \vdots & & \\ d_{J1} & \ldots & d_{JN} & d'_{J1} & \ldots & d'_{JN} \end{bmatrix},$$

$$d_{jn} = c_n \cos < \lambda_n, x_j >, \quad d'_{jn} = c_n \sin < \lambda_n, x_j >,$$

will be called *conditional spectral model* and will be denoted by $w_N^b(x)$. By Lemma 1.3 conditional spectral model minimizes the value

$$\sum_{j=1}^{J} [w_N(x_j) - b_j]^2$$

for fixed values of c_n, λ_n $(n = 1, \ldots, N)$.

3.4.1. Convergence of finite-dimensional distributions

By convergence of finite-dimensional distributions we understand weak convergence of corresponding vectors.

Lemma 3.7. *If finite-dimensional distributions of non-randomized spectral model $w_N(x)$ converge to finite-dimensional distributions of the Gaussian homogeneous field $w(x)$, then finite-dimensional distributions of conditional spectral model $w_N^b(x)$ converge to finite-dimensional distributions of the Gaussian random field $w^b(x)$.*

This result immediately follows from the assertion presented below, which can be easily proved in terms of characteristic functions. Suppose that a sequence of Gaussian random vectors

$$w_N = \begin{bmatrix} w_N^1 \\ w_N^2 \end{bmatrix} \in \mathbb{R}^m, \quad w_N^i \in \mathbb{R}^{m_i}, \quad m = m_1 + m_2,$$

weakly converges, as $N \to \infty$, to a Gaussian vector

$$w = \begin{bmatrix} w^1 \\ w^2 \end{bmatrix} \in \mathbb{R}^m, \quad w^i \in \mathbb{R}^{m_i},$$

and correlation matrix of the vector w^2 is a nonsingular one. Then conditional distributions of the vectors w_N^1 provided that $\|w_N^2 - b\| \to \min$ (b is an arbitrary vector) weakly converge to conditional distribution of the vector w^1 provided that $w^2 = b$.

Let us consider a sequence of partitionings of $\mathbb{R}^k$ into disjoint sets:

$$\Lambda_1^{(N)} + \ldots + \Lambda_N^{(N)} = \mathbb{R}^k, \tag{3.28}$$

(the superscripts are later omitted) and let us consider the spectral model (3.27) with

$$c_n^2 = \mu(\Lambda_n), \quad \lambda_n \in \Lambda_n. \tag{3.29}$$

The following condition

$$\Lambda_N \subset \{|\lambda| \geq d_N\}, \quad d \to \infty, \quad \max_{n<N} \operatorname{diam}(\Lambda_n) \underset{N\to\infty}{\to} 0 \tag{3.30}$$

is sufficient for convergence of finite-dimensional distributions of the spectral model (3.27)–(3.30) and, by Lemma 3.7, for convergence of finite-dimensional distributions of the corresponding conditional spectral model. In this case convergence occurs for an arbitrary algorithm of choice of $\lambda_j \in \Lambda_j$ including randomized spectral models as well. Thus, the following statement holds.

Lemma 3.8. *Finite-dimensional distributions of conditional spectral model* (3.27)–(3.30) *converge to the corresponding finite-dimensional distributions of the random field $w^b(x)$ (both for randomized and non-randomized models).*

3.4.2. Weak convergence in spaces L^p and C^p

The problem is particularly simple for weak convergence of randomized and non-randomized spectral models in spaces $L^p(K)$, K is a compact set in $\mathbb{R}^k$.

Lemma 3.9. *If conditional spectral model $w_N^b(x)$ converges to a Gaussian field $w^b(x)$ in the sense of convergence of finite-dimensional distributions, then it also converges in the sense of weak convergence in $L^p(K)$, $p \geq 1$.*

By Teorem 3.5, it is sufficient to prove that

$$\mathbf{M}\left|w_N^{(b)}(x)\right|^p \to \mathbf{M}\left|w^{(b)}(x)\right|^p \quad \text{as } N \to \infty \text{ and}$$

$$\sup_N \mathbf{M}\left|w_N^{(b)}(x)\right|^p < \infty.$$

But this is evident from convergence of one-dimensional distributions of the spectral model, which are Gaussian ones.

We proceed to weak convergence of conditional spectral models in the space $C^p(K)$.

As conditional spectral model $w_N^b(x)$ we first consider a non-randomized model that is constructed by (3.27)–(3.30) under the additional condition $|\lambda_N| = d_N$.

Lemma 3.10. *The conditional spectral model $w_N^b(x)$ weakly converges in $C^p(K)$, $p \in \{1, 2, \ldots\}$, to the Gaussian field $w^b(x)$ if*

$$\int |\lambda|^{\beta+2p}\mu(d\lambda) < \infty \tag{3.31}$$

for some $\beta > 0$.

Proof. 1. Consider the case $p = 0$. Convergence of finite-dimensional distributions follows from Lemma 3.7. We have to prove weak compactness of the sequence of random fields $w_N^b(x)$ in the space $C(K)$. For the field $w_N^b(x)$ we have:

$$\begin{aligned}\mathbf{M}\left|w_N^b(y) - w_N^b(x)\right|^2 =& B_N(x,x)+B_N(y,y)-2B_N(x,y)+\\ &+m_N^2(x) + m_N^2(y) - 2m_N(x)m_N(y) =\\ &=[B_N(x,x)-B_N(x,y)] + [B_N(y,y)-B_N(x,y)]+\\ &+[m_N(x) - m_N(y)]^2,\end{aligned}$$

where $B_N(x,y)$ is correlation function and $m_N(x)$ is expectation of the field $w_N^b(x)$. Let us consider matrices

$$B_N = \begin{bmatrix} B_N(x,x) & B_N(x,y) \\ B_N(x,y) & B_N(y,y) \end{bmatrix},$$

$$R_{(N)} = \begin{bmatrix} R_{(N)}(0) & R_{(N)}(x-y) \\ R_{(N)}(x-y) & R_{(N)}(0) \end{bmatrix},$$

$$K_{(N)} = \begin{bmatrix} R_{(N)}(x-x_1) & \dots & R_{(N)}(x-x_J) \\ R_{(N)}(y-x_1) & \dots & R_{(N)}(y-x_J) \end{bmatrix},$$

$$L_{(N)} = \left[R_{(N)}(x_i - x_j)\right]_{i,j=1,\dots,J}.$$

Here by $R_{(N)}(x)$ we denote correlation functions of homogeneous fields $w_{(N)}(x)$.

Matrices B_N and vectors

$$m_N = \begin{bmatrix} m_N(x) \\ m_N(y) \end{bmatrix}$$

satisfy the relations

$$B_N = R_N - K_N L_N^+ K_N^T, \quad m_N = K_N L_N^+ b.$$

From condition (3.31) it follows that for all N

$$R_N(0) - R_N(x) \le C\|x\|^\gamma,$$

where $\gamma = \min(2,\beta)$ and C is a constant (see, e.g. proof of Theorem 3.4). Since for correlation functions $R_N(x)$ we have

$$|R_N(x) - R_N(y)|^2 \le 2R_N(0)\,[R_N(0) - R_N(x-y)] \text{ and}$$

$$R_N \to R, \quad K_N \to K, \quad L_N^+ \to L^{-1} \text{ as } N \to \infty,$$

the following equalities hold for all N

$$|B_N(x,y) - B_N(x,x)| \le C'\|x-y\|^\gamma,$$

$$|m_N(x) - m_N(y)|^2 \le C'\|x-y\|^\gamma.$$

Hence, there exist the values of $\gamma \in (0,2]$ and $C'' > 0$ such that

$$\mathbf{M}\left|w_N^b(y) - w_N^b(x)\right|^2 \le C''\|x-y\|^\gamma \tag{3.32}$$

for all N. This inequality yields [52]

$$\mathbf{P}\left\{\sup_{x,y:\|x-y\|<h} \left|w_N^b(x) - w_N^b(y)\right| > a\right\} \le \varphi(h,a), \tag{3.33}$$

where $\varphi(h,a) \to 0$ as $h \to 0$. Since the right-hand side of (3.33) is independent of N, weak compactness of Gaussian random fields $w_N^b(x)$ in $C(K)$ is proved.

2. For the case $p > 0$ the proof of the lemma reduces to the proof of weak compactness of the fields $w_N^b(x)$ in $C^p(K)$. This in turn (see Theorem 3.3) reduces to checking relations of the form (3.32), (3.33) for partial derivatives of $w_N^b(x)$ of order p. These relations can be proved on the basis of (3.31), in the same way as for the case $p = 0$. This completes the proof of Lemma 3.10.

Along with spectral model (3.27) we consider a randomized spectral model

$$u_N(x) = \sum_{n=1}^{N-1} c_n \left[\xi_n \cos < \lambda_n, x > + \xi_n' \sin < \lambda_n, x >\right]. \qquad (3.34)$$

which satisfies (3.28)–(3.30). By $u_N^b(x)$ we denote the corresponding conditional spectral model constructed for (3.34).

Lemma 3.11. *The randomized conditional spectral model $u_N^b(x)$ weakly converges in $C^p(K)$, $p \in \{1, 2, \ldots\}$, to the Gaussian field $w^b(x)$ if*

$$\int |\lambda|^{\beta+2p} \mu(d\lambda) < \infty$$

for some $\beta > 0$.

The proof of Lemma 3.11 is analogous to that of Lemma 3.10. It is sufficient to emphasize that inequalities (3.32) and (3.33) hold for arbitrary vectors $\lambda_j \in \Lambda_j$, $j = 1, \ldots, N-1$, and, hence, for randomly chosen vectors as well.

Task 3.3. Obtain conditions for weak convergence of randomized models (3.27) in $C^p(K)$.

3.5. Remark on allowance for bias of estimates constructed by approximate models

Assume that we have to estimate $J = \mathbf{M}\xi$. This is a typical situation when modelling of ξ is difficult, but one can specify a comparatively simple algorithm for simulation of a sequence of random variables $\xi(m)$ converging to ξ. Henceforth it will be assumed that

$$\sup_m \mathbf{M}|\xi(m)|^{1+\beta} < \infty \quad \text{for some } \beta > 0.$$

This ensures the convergence

$$J(m) = \mathbf{M}\xi(m) \to J.$$

Thus, as estimates of J, we can use asymptotically unbiased estimates

$$J(m, N) = \sum_{i=1}^{N} \xi_i(m)/N,$$

where $\xi_i(m)$ are independent realizations of $\xi(m)$.

To estimate the bias $J(m) - J$ and to take it into account, we may take advantage of the Runge method and the Aitken process. Let

$$J(m) = J + am^{-p} + O(m^{-q}), \quad q > p > 0. \tag{3.35}$$

Then accurate to $O(m^{-q})$, we have

$$J(m, N) \approx J + am^{-q} + \sigma(m)\varepsilon N^{-1/2},$$

where ε is a standard normal variable and $\sigma^2(m)$ is a variance of $\xi(m)$. Assume that the value of p is known. Consider the following estimate

$$\begin{aligned} J^* &= \frac{-J_1 m_2^{-p}}{m_1^{-p} - m_2^{-p}} + \frac{J_2 m_1^{-p}}{m_1^{-p} - m_2^{-p}}, \\ J_i &= J(m_i, N_i), \qquad i = 1, 2. \end{aligned} \tag{3.36}$$

Bias of J^* is of order m_i^{-q}:

$$\begin{aligned} J^* &= J - \sigma_1\varepsilon_1 N_1^{-1/2} m_2^{-p} / \left(m_1^{-p} - m_2^{-p}\right) + \\ &+ \sigma_2\varepsilon_2 N_2^{-1/2} m_1^{-p} / \left(m_1^{-p} - m_2^{-p}\right) + \\ &+ O(m_1^{-q}) + O(m_2^{-q}), \quad \sigma_i = \sigma(m_i). \end{aligned}$$

Variance of J^* equals

$$\mathbf{D}J^* = \left(\sigma_1^2 m_2^{-2p} N_1^{-1} + \sigma_2 m_1^{-2p} N_2^{-1}\right) / \left(m_1^{-p} - m_2^{-p}\right)^2 \tag{3.37}$$

provided that J_1 and J_2 are independent. The use of estimate J^* is reasonable when biases of J_i are significant in comparison with statistical error so that value of a is well estimated by formula

$$a \approx (J_1 - J_2)/(m_1^{-p} - m_2^{-p}). \tag{3.38}$$

Remark. In choosing N_1, N_2 one can use a relation which results from minimality condition for DJ^* in (3.37) with restriction $t(m_1)N_1 + t(m_2)N_2 = T$, where $t(m)$ is time of modelling $\xi(m)$. If $t(m) = bm$, then optimal values of N_j, $j = 1, 2$, satisfy the relation

$$\sigma_1^2 m_1^{2p-1} N_1^{-2} = \sigma_2 m_2^{2p-1} N_2^{-2}.$$

Minimizing the variance in (3.38) leads to

$$N_1 m_1^{-1/2}/\sigma_1 = N_2 m_2^{-1/2}/\sigma_2.$$

If the value of p is unknown, then it can be estimated in terms of

$$p \approx \ln B / \ln \rho, \quad \rho > 1, \tag{3.39}$$
$$B = [J(m, N_1) - J(\rho m, N_2)] \,/\, \left[J(\rho m, N_2) - J(\rho^2 m, N_3)\right],$$

where N_1, N_2, N_3 are large enough.

Task 3.4. Study statistical properties of the estimate (3.39) for p.

Estimates (3.36)–(3.39) repeat formulas of the Runge method and the Aitken process (see, for example, [108]) while the requirement of a small effect of statistical error makes them time-consuming. The situation can be somewhat improved by constructing dependent estimates of $J(m_j, N_j)$.

Note that approximate formulas (3.38)–(3.39) can be used to find optimal relations between m and N while constructing additional estimates $J(m, N)$. Let $\sigma(m)$ be independent of m: $\sigma(m) = \sigma$. Then minimization of

$$\mathbf{M}\,[J(m, N) - J]^2 \approx a^2 m^{-2p} + \sigma^2 N^{-1}$$

under the constraint $Nt(m) = T$, determines an optimal relation between m and N. For $t(m) \sim m$ we obtain

$$N = m^{2p} \sigma^2 / (2pa^2).$$

Let us consider a model problem of determining an average number of crossings of the level $w = 0$ on the segment $x \in [0, 1]$ by a stationary Gaussian process $w(x)$ with spectral density

$$S(\lambda) = \begin{cases} 1/(2C), & |\lambda| \le C, \\ 0, & |\lambda| > C. \end{cases}$$

Numerical solution of the problem was compared to the exact solution

$$J = 3^{-1/2} C \exp(-H^2/2)/\pi \approx 18.38 \tag{3.40}$$

for $C = 100$ and $H = 0$. For approximate modelling of the process $w(x)$ we used spectral model

$$w_m(x) = m^{-1/2} \sum_{i=1}^{m} (-2 \ln \alpha_j)^{1/2} \cos(\lambda_i x + 2\pi \alpha_j'),$$
$$\lambda_j = 100j/m,$$

where α_j, α_j' are independent random variables uniformly distributed over the segment (0,1). Using the spectral model we constructed the estimates

$$J(m, N) = \sum_{i=1}^{N} f\left(w_m^{(i)}\right)/N,$$

where $w_m^{(i)}$ are independent trajectories of the processes w_m and f is number of zero crossings. The results of Section 3.3 provide convergence

$$J_m = \mathbf{M} f(w_m) \to J.$$

Below we present results of computations for the model problem making use of estimates (3.39), (3.36):

$$J(2, 1000) = 24.85, \quad J(4, 1000) = 21.56, \quad J(8, 1000) = 19.94,$$
$$p \approx 1.022, \quad J^* = 18.37.$$

In this case, the value of J^* was practically the same as the exact value (3.40).

Obviously, applying such a procedure is reasonable when bias decreases slowly, and time of modelling for $\xi(m)$ rapidly increases with m.

Chapter 4

On optimization and convergence of functional Monte Carlo estimators

The Chapter deals with convergence and optimization of unbiased functional Monte Carlo estimators. We shall obtain estimators, which are optimal in Sobolev's Hilbert spaces, for computing integrals depending on a parameter and for computing families of functionals of solution of integral equations of the second kind. The results will be obtained in the context of a new concept proposed by the author in order to compare the efficiency of functional estimators in Monte Carlo method. For weak convergence of functional Monte Carlo estimators we shall prove sufficient conditions suitable for various applications.

In the Chapter we use the following notation: $\mathcal{D}$ is operation of differentiation; $\|f\|$ is the norm of function $f(x)$ in a Sobolev's Hilbert space H; if $f_\omega = f(x,\omega)$ is a family of functions, then $\|f(.,\omega)\|_H$ are the corresponding values of the norms.

4.1. Asymptotic efficiency of estimators in the method of dependent tests. Convergence conditions

4.1.1. Asymptotic computational cost

Monte Carlo method for calculating a quantity J consists in constructing a random variable ξ (Monte Carlo estimator) such that $\mathbf{M}\xi = J$ and in computing the estimate

$$J_n^* = \sum_{i=1}^{n} \xi_i/n, \tag{4.1}$$

where ξ_i are independent realizations of ξ. In case of finite variance, the value

$$S = nt\mathbf{M}|J_n^* - J|^2 = nt\mathbf{D}J_n^* = t\mathbf{D}\xi, \tag{4.2}$$

is called *computational cost* of the estimate J_n^* (of the estimator ξ). Here t is computing time of a single realization of ξ.

We now consider a case where Monte Carlo method is employed to calculate a function $J(x)$. Let $\xi(x)$ be a random function such that

$$J(x) = \mathbf{M}\xi(x).$$

Assume that an estimate $J_n^*(x)$ of the function $J(x)$ is constructed by the *method of "dependent tests"*:

$$J_n^*(x) = \sum_{i=1}^{n} \xi_i(x)/n, \tag{4.3}$$

where $\xi_i(x)$ are independent realizations of $\xi(x)$. (The conventional term 'dependent tests' implies that the values of $\xi(x)$ may be dependent statistically for different x). The method of dependent tests is widely employed in Monte Carlo computations for integrals depending on a parameter; for families of functionals of solutions of integral equations; for solving partial differential equations; for simulation of radiation transfer processes (the method of similar trajectories), and so on (see, e.g. [22], [28], [111].)

In paper [90] the following generalization of the concept of the computational cost has been suggested for a functional case. We suppose that the function $J(x)$ and realizations of $\xi(x)$ belong to a normed space F. The quantity

$$S_n = n\,t\mathbf{M}\|J_n^* - J\|_F^2 = t\mathbf{M}\left\|\sum_{i=1}^{n} [\xi_i(x) - J(x)]\right\|_F^2 /n, \tag{4.4}$$

will be called the *computational cost* of the estimate J_n^* (in the space F), where t is computing time of a single realization of $\xi(x)$, and $\|.\|_F$ is a norm in the space F. Under some additional assumptions considered below we have

$$n\mathbf{M}\|J_n^* - J\|_F^2 \to \mathbf{M}\|w\|_F^2 \quad \text{as } n \to \infty, \tag{4.5}$$

where $w(x)$ is a Gaussian stochastic process with mean zero and covariance function

$$\mathbf{M}w(x)w(y) = \mathbf{M}\,[\xi(x) - J(x)]\,[\xi(y) - J(y)]\,. \tag{4.6}$$

Condition (4.5) allows to introduce the concept of *asymptotic computational cost* of the estimates J_n^*:

$$S = t\mathbf{M}\|w\|_F^2 = \lim_n S_n. \tag{4.7}$$

Thus, the quantity $\mathbf{M}\|w\|_F^2$, which will be later called *F-deviation* of the estimator ξ, and denoted by

$$\mathrm{Dev}(\xi, F),$$

characterizes the efficiency of the estimate (4.3) in the space F. The estimator ξ (from a class of admissible estimators), for which minimal F-deviation is attained is called F-*optimal* estimator (in this class).

Remark 1. Let us justify once more the choice of quantity (4.7) as asymptotic computational cost. The term computational cost of a Monte Carlo estimator usually implies computing time required to attain the necessary accuracy. As an error of the functional estimate we consider $(\mathbf{M}\|J_n^* - J\|_F^2)^{1/2}$. The time that is necessary to simulate $J(x)$ with given accuracy we denote by $T(\Delta)$:

$$T(\Delta) = t\,n(\Delta), \quad n(\Delta) = \min\left\{n : \sqrt{\mathbf{M}\|J_n^* - J\|_F^2} \le \Delta\right\}.$$

Since (4.5) we have $T(\Delta)\Delta^2 \sim t\mathbf{M}\|w\|_F^2$ for small Δ.

Remark 2. If we use $(\mathbf{M}\|J_n^* - J\|_F^2)^{1/p}$, $p > 0$, as an error of functional estimate J_n^*, then under assumption

$$\mathbf{M}\|n^{-1/2}(J_n^* - J)\|_F^p \to \mathbf{M}\|w\|_F^p$$

the following is fulfilled

$$T(\Delta)\Delta^2 \sim t\,(\mathbf{M}\|w\|_F^p)^{2/p},$$

and $(\mathbf{M}\|w\|_F^p)^{2/p}$ can be used as a measure of asymptotic efficiency of the estimate J_n^*.

4.1.2. Convergence conditions

Below we present conditions that ensure convergence (4.5), (4.7) in some functional spaces F. We denote

$$\varepsilon(x) = \xi(x) - J(x).$$

Statement 4.1. *Assume that realizations of $\varepsilon(x)$ belong to a normed space F. If random variables*

$$n\|J_n^* - J\|_F^2 = n^{-1}\left\|\sum_{i=1}^{n} \varepsilon_i\right\|_F^2$$

(here ε_i are independent realizations of ε) are uniformly integrable, then the conditions presented below are sufficient for existence of the limit (4.7) and convergence (4.5):

(a) *For the space $F = L^p(X, \nu)$, where $p \geq 2$ and (X, ν) is a measurable space with a σ – finite measure ν, sufficient convergence condition is*

$$\mathbf{M}\|\varepsilon\|_F^2 < \infty.$$

(b) *For the space $F = C(X)$, where X is a compactum in $\mathbb{R}^k$, sufficient convergence conditions are:*

$$\mathbf{M}\varepsilon^2(x) < \infty \quad \text{for all } x \text{ from } X,$$

and there exist $\alpha > 0$ and a random variable $G(\omega) > 0$, $\mathbf{M}G^2 < \infty$, such that

$$|\varepsilon(x, \omega) - \varepsilon(y, \omega)| < G(\omega)|x - y|^\alpha$$

for all x and y from X.

(c) *For the space $F = C^p(X)$, where $p \in \{1, 2, \ldots\}$ and X is a bounded closed domain in $\mathbb{R}^k$, sufficient convergence conditions are:*

$$\max_{|m| \leq p} \sup_{x \in X} \mathbf{M}\left[\mathcal{D}^m \varepsilon(x)\right]^2 < \infty,$$

and for any multi-index $m = (m_1, \ldots, m_k)$ such that $|m| = p$ there exist a random variable $G_m(\omega) > 0$, $\mathbf{M}G_m^2 < \infty$, and a number $\alpha_m > 0$, for which

$$\left|\mathcal{D}^m\left[\varepsilon(x, \omega) - \varepsilon(y, \omega)\right]\right| < G_m(\omega)|x - y|^{\alpha_m}, \quad x, y \in X.$$

Here

$$m = (m_1, m_2, \dots, m_k), \quad |m| = m_1 + \dots + m_k,$$

and by

$$\mathcal{D}^m \dots = \frac{\partial}{\partial x_1^{m_1} \partial x_2^{m_2} \dots \partial x_k^{m_k}} \dots$$

we denote partial derivatives of trajectories of a random field.

(d) *In a separable Hilbert space F sufficient convergence condition is:*

$$\mathbf{M}\|\varepsilon\|_F^2 < \infty.$$

Proof. Conditions (a)–(d) imply weak convergence in the space F for the random functions $n^{1/2}(J_n^*(x) - J(x))$ to the Gaussian process $w(x)$ with covariance function (4.6) (statements of the corresponding functional limit theorems can be found in [45], [87], [89], [93], [100]). Since the functional $\|.\|_F^2$ is continuous in F, we have weak convergence of $n\|J_n^* - J\|_F^2$ to $\|w\|_F^2$, and the uniform integrability of the quantities $n\|J_n^* - J\|_F^2$ ensures convergence of the moments in (4.5), (4.7) (cf. [13]).

Remark 1. Let us recall the definition of weak convergence. A sequence of random elements ξ_n is said to converge weakly to a random element ξ in the space F with metric ρ_F if for any real functional g, continuous in the metric ρ_F, the sequence of random variables $g(\xi_n)$ converges to random variable $g(\xi)$ in distribution.

Because of the weak convergence of $n^{1/2}(J_n^*(x) - J(x))$ to the Gaussian process $w(x)$ in the space F we can state that the deviation of J_n^* from J on X is of the order $n^{-1/2}$ in probability:

$$\mathbf{P}\left\{\rho_F(J_n^*, J) \le c n^{-1/2}\right\} \to \mathbf{P}\left\{\rho_F(\omega, 0) \le c\right\}, \quad n \to \infty.$$

Therefore, the conditions for weak convergence of $n^{-1/2}(J_n^*(x) - J(x))$ to $w(x)$ in various functional spaces are of separate interest for general theory of Monte Carlo methods. The conditions (a)-(d) of Statement 4.1 are general conditions which are sufficient for weak convergence in corresponding spaces. On the basis of these conditions, one can obtain efficient criteria for weak convergence of various functional estimates that are used in Monte Carlo methods (see Item 4.3 below).

Remark 2. The uniform integrability of the quantities $n\|J_n^* - J\|_F^2$ in Statement 4.1 can be replaced by the following condition

$$\sup_n n^{-1-\beta}\mathbf{M}\Big\|\sum_{i=1}^{n}\varepsilon_i(x)\Big\|_F^{2+2\beta} < \infty,$$

where $\beta > 0$, and ε_i are independent realizations of $\varepsilon = \xi - J$.

Remark 3. Conventionally, in order to compare the quality of unbiased functional Monte Carlo estimators one uses *the criterion of weighted variance* [22] based on comparison of the quantities

$$\int_X \mathbf{D}\xi(x)\nu(dx),$$

or *the minimax criterion* proposed and studied by G.A.Milchailov [74] based on the comparison of maxima of variances

$$\max_{x\in X} \mathbf{D}\xi(x).$$

Without analyzing in detail various criteria for optimality we note that a new approach described here is quite reasonable and in general case it is not reduced to minimax criterion or criterion of weighted variance. The idea is that for comparing the quality of estimators we use the quantities

$$n\mathbf{M}\|J_n^* - J\|_F^2 = \mathbf{M}\Big\|\sum_{i=1}^{n}[\xi_i(x) - J(x)]\Big\|_F^2/n,$$

and their limit value

$$\mathrm{Dev}(\xi, F) = \mathbf{M}\|w\|_F^2 \tag{4.8}$$

The quantity (4.8) is essentially a functional of the correlation function (4.6):

$$\mathrm{Dev}(\xi, F) = \Phi(R(x,y)),$$
$$R(x,y) = \mathbf{M}\,[\xi(x) - J(x)]\,[\xi(y) - J(y)]\,.$$

Thus, the problem to get analytical representations of functional Φ for various spaces F is an urgent one.

Remark 4. For the spaces $F = L^2(X,\nu)$ the equality

$$n\mathbf{M}\|J_n^* - J\|_F^2 = \mathbf{M}\|w\|_F^2 = \int_X \mathbf{D}\xi(x)\nu(dx),$$

is valid and the problem is reduced to the well-known problem of minimization of weighted variance. An analogous equality is valid for Sobolev's Hilbert spaces. This enables us to obtain the results which can be considered as a generalization of the method of "importance sampling" (about "importance sampling" see, for example, [22], [23], [74]). The next section is devoted to such generalization.

4.2. Optimal functional estimators in Sobolev's Hilbert spaces

4.2.1. *F*-deviation of estimators in Sobolev's Hilbert spaces

Let X be a bounded domain in $\mathbb{R}^k$ and $C^p(X)$ be the space of functions on X with continuous partial derivatives of the order less or equal to p, $p \in \{0,1,2,\ldots\}$. We suppose that a measure μ_m is specified on X for each multi-index $m = (m_1,\ldots,m_k)$ such that $0 \le |m| = m_1+\ldots+m_k \le p$. Having demanded that the measure $\mu_0 = \mu(0,\ldots,0)$ be equivalent to the Lebesgue measure we consider the scalar product in $C^p(X)$

$$< f,g >= \sum_{|m|\le p} \int_X \mathcal{D}^m f(x)\mathcal{D}^m g(x)\mu_m(dx), \tag{4.9}$$

$$\mathcal{D}^m f(x) = \frac{\partial^{|m|}}{\partial x_1^{m_1}\ldots\partial x_k^{m_k}} f(x).$$

By $H = H^p(X,\{\mu_m\})$ we denote the completion of the space $C^p(X)$ in the norm

$$\|f\|_H = \left[\sum_{|m|\le p} \int_X |\mathcal{D}^m f(x)|^2 \mu_m(dx)\right]^{1/2}. \tag{4.10}$$

The space H is a Hilbert space with the scalar product (4.9) where $\mathcal{D}^m f(x)$ and $\mathcal{D}^m g(x)$ are the corresponding generalized derivatives.

Henceforth we assume that

(A) the realizations of $\xi(x)$ and the function $J(x)$ belong to the space H;

(B) $\mathbf{M}\mathcal{D}^m\xi(x) = \mathcal{D}^m J(x)$ (i. e. $\mathbf{M}\mathcal{D}^m\varepsilon(x) = 0$) for $0 \leq |m| \leq p$,

(C) $\int\limits_X \mathbf{M}|\mathcal{D}^m\varepsilon(x)|^2\mu_m(dx) < \infty$, $0 \leq |m| \leq p$.

It is easy to verify that under these conditions the following equalities

$$\begin{aligned}\mathrm{Dev}(\xi, H) &= n\mathbf{M}\|J_n^* - J\|_H^2 = \mathbf{M}\|\xi - J\|_H^2 = \\ &= \mathbf{M}\|\xi\|_H^2 - \|J\|_H^2 = \mathbf{M}\|\varepsilon\|_H^2 = \mathbf{M}\|w\|_H^2 \end{aligned} \tag{4.11}$$

hold for H-deviation of ξ, and the computational cost of the estimate J_n^* does not depend on n: $S_n \equiv S$.

The criterion for optimality of Monte Carlo estimators which is based on the H-deviation naturally arises while calculating simultaneously a function and its derivatives by the method of dependent tests.

4.2.2. H-optimal estimators for computing integrals depending on a parameter

In order to calculate an integral depending on the parameter x:

$$J(x) = \int\limits_\Omega G(x,\omega)d\nu(\omega), \quad x \in X,$$

we consider the family of estimates

$$\begin{aligned} J_n^*(x) &= n^{-1}\sum_{i=1}^n \xi_i(x), \\ \xi(x,\omega) &= G(x,\omega)d\nu/d\lambda(\omega), \end{aligned} \tag{4.12}$$

where $\xi_i(x) = \xi(x,\omega_i)$ are independent realizations of the random function $\xi(x,\omega)$, ω is a random variable distributed according to probability measure $d\lambda$, measure $d\nu$ is absolutely continuous with respect to $d\lambda$, and $d\nu/d\lambda$ is the density of measure $d\nu$ with respect to $d\lambda$.

Under assumptions (A)–(C) the following lemma is valid.

Lemma 4.1. *The H-optimal estimator in the class* (4.12) *exists and is unique (accurate to equivalence with respect to measure $d\nu$). The probability measure $d\lambda$ that minimizes H-deviation* (4.11) *of the estimator* (4.12) *is defined by relation*

$$d\hat{\lambda}(\omega) = \|G(.,\omega)\|_H d\nu(\omega) / \int_\Omega \|G(.,\omega)\|_H d\nu(\omega). \tag{4.13}$$

Proof. According to (4.11) minimization of H-deviation is reduced to minimization of $\mathbf{M}\|\xi\|_H^2$ for which we have

$$\mathbf{M}\|\xi\|_H^2 \geq (\mathbf{M}\|\xi\|_H)^2. \tag{4.14}$$

Since

$$\xi(x) = G(x,\omega) d\nu / d\lambda(\omega),$$
$$\mathbf{M}\|\xi\|_H^2 = \int_\Omega \|G(.,\omega)\|_H^2 d\nu / d\lambda(\omega) d\nu(\omega),$$
$$\mathbf{M}\|\xi\|_H = \int_\Omega \|G(.,\omega)\|_H d\nu(\omega),$$

the equality in (4.14) is reached if and only if

$$d\nu / d\lambda(\omega) = \mathbf{M}\|\xi\|_H / \|G(.,\omega)\|_H.$$

Hence

$$d\hat{\lambda}(\omega) = \|G(.,\omega)\|_H d\nu(\omega) / \mathbf{M}\|\xi\|_H.$$

Lemma has been proved.

Example. We consider

$$J(x) = \int_0^\infty x \exp(-\omega x) d\omega \equiv 1, \quad x \in [a,b],$$

as an integral depending on a parameter, and we take the spaces

$$H_0 = L^2[a,b], \quad H_1 = H_1[a,b], \quad H_2 = H_2[a,b], \quad C = C[a,b],$$

as the functional space F. Assume that the distributions μ_0, μ_1 and μ_2 which are used in definition of the spaces H_0, H_1 and H_2 are uniform probability distributions in $[a, b]$. By ξ_0, ξ_1, ξ_2 we denote F-optimal estimators for corresponding spaces $F = H_0, H_1, H_2$, and by $\bar{\xi}$ we denote the minimax estimator. The results of computations of F-deviations are presented in the next Table for $a = 1$, $b = 3$ (these results are borrowed from [77]).

deviation	ξ_0	ξ_1	ξ_2	$\bar{\xi}$
H_0-deviation	0.17	0.24	0.45	0.20
H_1-deviation	1.0	0.78	0.94	0.84
H_2-deviation	5.6	2.23	1.64	6.39
C-deviation	0.37	0.18	0.29	0.24

In this example the H_1-optimal estimator turns out to be the best one of the four estimators in terms of C-deviation criterion.

We emphasize here that in order to calculate C-deviation we don't need to simulate the corresponding estimators, which may turn out to be a time-consuming procedure. It is sufficient to find correlation function of estimator and then compute (by Monte Carlo method) the expectation of the squared C-norm of a Gaussian process with zero mean and the obtained correlation function.

Remark. For the estimate (4.1) the problems of minimization of absolute and relative errors coincide, while in the functional case the question arises: What should be chosen as a measure of relative error? One of possible approaches is as follows. We assume that $J(x) \neq 0$ for $x \in X$. The value

$$\hat{S}_n = nt\mathbf{M}\|(J_n^* - J)/J\|_F^2 = t\mathbf{M}\left\| n^{-1/2} \sum_{i=1}^{n} \varepsilon_i / J \right\|_F^2$$

is said to be the relative computational cost of the estimate J_n^* in the space F.

Under additional assumptions analogous to those for the convergence (4.5), (4.7) we have

$$\hat{S}_n \to \hat{S}_\infty = t\mathbf{M}\|w(x)/J(x)\|_F^2, \quad \text{as } n \to \infty,$$

where $w(x)/J(x)$ is the corresponding Gaussian function.

In the spaces $F = H$ the relative computational cost of the estimates J_n^* (under conditions analogous to (A)–(C)) is independent of n:

$$\hat{S}_n \equiv t\mathbf{M}\|\varepsilon/J\|_H^2 = t\left[\mathbf{M}\|\xi/J\|_H^2 - C\right],$$
$$C = \int_X \mu_0(dx),$$

and, as in Lemma 4.1, one can prove a statement about minimum of mean squared norm of the relative error

$$\mathbf{M}\|\varepsilon/J\|_H^2$$

of the estimator (4.12) for an integral depending on a parameter.

4.2.3. H-optimal "absorption" estimator for computing a family of functionals of solution of an integral equation of the second kind

In what follows we use the notation and terminology from [22], [74].

Let us consider function

$$J(x) = (\varphi, h(x)) = \int_\Lambda \varphi(\lambda)h(x,\lambda)d\lambda, \quad x \in X,$$

where Λ is a measurable set in $\mathbb{R}^k$; the family $h(x) = h(x,\lambda)$, $x \in X$, belongs to the space of substantially bounded functions $L^\infty = L^\infty(\Lambda)$ with the norm

$$\|h\|_\infty = \operatorname*{vrai\ sup}_{\lambda\in\Lambda} |h(\lambda)|,$$

and $\varphi(\lambda)$ is solution of integral equation of the second kind

$$\varphi(\lambda) = \int_\Lambda K(\lambda',\lambda)\varphi(\lambda')d\lambda' + f(\lambda),$$

or in the operator form:

$$\varphi = K\varphi + f,$$

$$\varphi, f \in L^1 = L^1(\Lambda), \quad K : L^1 \to L^1.$$

We denote by K^* the operator adjoint to the operator K and acting in L^∞. Note that

$$J(x) = (\varphi, h(x)) = (f, \psi(x)),$$

where $\psi(x)$ is a solution of the adjoint equation in L^∞:

$$\psi(x) = K^*\psi(x) + h(x),$$

i.e.

$$\psi(x, \lambda) = \int_\Lambda K(\lambda, \lambda')\psi(x, \lambda')d\lambda' + h(x, \lambda).$$

In order to calculate $J(x)$ we consider the *"absorption" estimator* (cf. [22], [74]):

$$J_n^*(x) = n^{-1}\sum_{i=1}^{n} \xi_i(x),$$

$$\xi(x) = Q_N h(x, \lambda_N)/g(\lambda_N). \tag{4.15}$$

Here $\lambda_0, \lambda_1, \lambda_2, \ldots$ is a homogeneous Markov chain with initial density $\pi(\lambda_0)$, transition density $p(\lambda', \lambda)$, and extinction probability $g(\lambda') = 1 - \int_\Lambda p(\lambda', \lambda)d\lambda$ at the point λ'; N is the random number of the last state;

$$Q_0 = f(\lambda_0)/\pi(\lambda_0), \quad Q_n = Q_{n-1}K(\lambda_{n-1}, \lambda_n)/p(\lambda_{n-1}, \lambda_n).$$

It is well-known [22] that, if the following conditions are satisfied

(a) $\pi(\lambda) \neq 0$ for $f(\lambda) \neq 0$,
(b) $p(\lambda', \lambda) \neq 0$ for $K(\lambda', \lambda) \neq 0$,
(c) $g(\lambda) \neq 0$, if $h(x, \lambda) \neq 0$ for a certain $x \in X$,
(d) *spectral radius* $\mathrm{Rad}(K_1) = \lim \|K_1^n\|^{1/n}$ is less than unit, where K_1 is an integral operator with the kernel

$$K_1(\lambda',\lambda) = |K(\lambda',\lambda)|,$$

then $\mathbf{M}\xi(x) = J(x)$.

If Monte Carlo algorithms for solving integral equations of the second kind are based on optimal (in a sense) estimators or estimators close to the optimal, then the term *the "importance sampling" technique* is used and $\psi(x,\lambda)$ is called the *importance function* for the point λ with respect to the functional J(x) (see [22], [74]). We obtain below H-optimal absorption estimators, i.e. the estimators of the form (4.15) which minimize the mean squared norm of the error in Sobolev's Hilbert spaces:

$$\mathrm{Dev}(\xi, H) = \mathbf{M}\|\varepsilon\|_H^2 \to \min,$$
$$\varepsilon(x) = \xi(x) - J(x).$$

A generalization of familiar result concerning the importance sampling technique for the absorption estimator for many functionals is formulated as follows.

Theorem 4.1. *Assume that condition (d) is satisfied, and the strictly positive function*

$$h_0(\lambda) = \|h(.,\lambda)\|_H = \left[\sum_{|m|\le p} \int_X |\mathcal{D}^m h(x,\lambda)|^2 \mu_m(dx)\right]^{1/2},$$

which belongs to the space L^∞, is defined (recall that $\mathcal{D}^m$ denotes an operator of differentiation with respect to the parameter $x \in X \subset \mathbb{R}^k$). Denote by ψ_0 and $\bar{\varphi}$ the solutions of the integral equations

$$\psi_0 = K_1^* \psi_0 + h_0,$$
$$\bar{\varphi} = K_1 \bar{\varphi} + |f|.$$

Then the Markov chain for the H-optimal absorption estimator $\xi(x)$ from (4.15) is defined by the formulas

$$\pi(\lambda) = |f(\lambda)|\psi_0(\lambda)/(|f|,\psi_0),$$
$$p(\lambda',\lambda) = |K(\lambda',\lambda)|\psi_0(\lambda)/\psi_0(\lambda'), \tag{4.16}$$
$$g(\lambda) = h_0(\lambda)/\psi_0(\lambda).$$

Moreover, for the H-optimal absorption estimator we have

$$(\mathbf{M}\|\xi\|_H)^2 = \mathbf{M}\|\xi\|_H^2 = (|f|, \psi_0)^2 = (\bar{\varphi}, h_0)^2.$$

Proof. By virtue of the assumptions (A)–(C) (see Item 4.2.1) the minimization of the H-deviation is reduced to minimization of the quantity

$$\begin{aligned} \mathbf{M}\|\xi\|_H^2 &= \mathbf{M}\|Q_N h(., \lambda_N)/g(\lambda_N)\|_H^2 = \\ &= \mathbf{M}\left[Q_N \|h(., \lambda_N)\|_H / g(\lambda_N)\right]^2 = \mathbf{M}\xi_0^2, \\ \xi_0 &= |Q_N| h_0(\lambda_N)/g(\lambda_N). \end{aligned}$$

The random variable ξ_0 is the absorption estimator for the quantity $(|f|, \psi_0) = (\bar{\varphi}, h_0)$. It is easy to verify that for the Markov chain defined by (4.16) the estimator ξ_0 is unbiased,

$$\mathbf{M}\xi_0 = (|f|, \psi_0) = (\bar{\varphi}, h_0),$$

and

$$\mathbf{M}\xi_0^2 = (|f|, \psi_0)^2 = (\bar{\varphi}, h_0)^2.$$

Thus, the relations (4.16) make the estimator $\xi_0 = \|\xi\|_H$ degenerate and give the minimum of $\mathbf{M}\|\xi\|_H^2$. The Theorem has been proved.

Exercise 4.1. Prove the uniqueness of the H-optimal absorption estimator (4.15) (see also, Theorem 6.4 in [21]).

Remark 1. For the absorption estimate the following is fulfilled

$$\mathbf{M}\|\xi\|_F^2 = \mathbf{M}\xi_0^2, \quad \xi_0 = \|\xi\|_F = Q_N \frac{\|h(., \lambda_N)\|_F}{g(\lambda_N)},$$

$$\mathbf{M}\xi_0^2 = (\chi, \|h\|_F^2/g) = (f^2/\pi, \chi^*),$$

where χ, χ^* are the Neumann series:

$$\chi = \sum_{n=0}^{\infty} L^n \frac{f^2}{\pi}, \quad \chi^* = \sum_{n=0}^{\infty} (L^*)^n \frac{\|h\|_F^2}{g}$$

for the integral equations of the second kind

$$\begin{aligned} &\chi = L\chi + f^2/\pi, \quad \chi^* = L^*\chi^* + \|h\|_F^2/g, \\ &L\chi(\lambda) = \int_\Lambda K^2(\lambda', \lambda)/p(\lambda', \lambda)\chi(\lambda')d\lambda'. \end{aligned}$$

Remark 2. The problem of obtaining an H-optimal *"collision" estimator*

$$\xi(x) = \sum_{n=0}^{N} Q_n h(x, \lambda_n)$$

is a more complicated one. Note that if

$$\xi_0 = \sum_{n=0}^{N} |Q_n| \, \|h(.,\lambda_n)\|_F,$$

then, evidently,

$$\|\xi\|_F \leq \xi_0,$$

and hence,

$$\mathbf{M}\|\xi\|_F^2 \leq \mathbf{M}\xi_0^2.$$

This inequality gives grounds to use an approximate information on the importance function with respect to the functional $< \bar{\varphi}, \|h\|_F >= \mathbf{M}\xi_0$ in order to optimize the collision estimate $\xi(x)$.

4.2.4. Investigation of H-optimal "collision" estimator

The main principle of investigating H-optimal collision estimator consists in reducing the problem to the known results for the non-functional case (as it was done for absorption estimator). Therefore, we need the following assertion.

Lemma 4.2. *Let us consider an integral equation of the second kind*

$$\varphi(\lambda) = \int_{\Lambda} K(\lambda', \lambda)\varphi(\lambda')d\lambda' + f(\lambda), \quad \varphi = K\varphi + f,$$

$$\varphi, f \in L^1 = L^1(\Lambda), \quad K : L^1 \to L^1,$$

and the collision estimator

$$\xi_1 = \sum_{n=0}^{N} Q_n h_1(\lambda_n) \tag{4.17}$$

for

$$J_1 = (\varphi, h_1) = \int_\Lambda \varphi(\lambda) h_1(\lambda) d\lambda = (f, \psi_1), \tag{4.18}$$

where

$$\psi_1 = K^* \psi_1 + h_1, \quad \psi_1, h_1 \in L^\infty = L^\infty(\Lambda), \quad K^* : L^\infty \to L^\infty.$$

I. *If the conditions are satisfied:*

$$\begin{aligned} &\text{a)} \quad \pi(\lambda) \neq 0 \quad \text{for } f(\lambda) \neq 0, \\ &\text{b)} \quad p(\lambda', \lambda) \neq 0 \quad \text{for } K(\lambda', \lambda) \neq 0, \\ &\text{c)} \quad \mathrm{Rad}(K_1) = \lim \|K_1^n\|^{1/n} < 1, \end{aligned}$$

where K_1 *is the integral operator with the kernel*

$$K_1(\lambda', \lambda) = |K(\lambda', \lambda)|,$$

then $\mathbf{M}\xi_1 = J_1$.

II. *Under comparatively general conditions, for the second moment of the random variable* ξ_1 *the relations hold (for more details, see* [21], [22], [70]*):*

$$\mathbf{M}\xi_1^2 = (\chi, 2h_1\psi_1 - h_1^2) = (f^2/\pi, \chi_1^*),$$

where

$$\chi = L\chi + f^2/\pi, \quad \chi_1^* = L^*\chi_1^* + h_1(2\psi_1 - h_1),$$

and L *is the integral operator with the kernel* $K^2(\lambda', \lambda)/p(\lambda', \lambda)$,

$$L\chi(\lambda) = \int_\Lambda K^2(\lambda', \lambda)/p(\lambda', \lambda)\chi(\lambda') d\lambda'.$$

III. *If* $\mathrm{Rad}(K_1) < 1$, *there exists a unique Markov chain* λ_n *for which the collision estimator* ξ_1 *(4.17) has a minimal variance. The initial and transition densities for this Markov chain are*

$$\pi(\lambda) = |f(\lambda)|\rho(\lambda)/(|f|, \rho),$$

$$p(\lambda, \lambda') = |K(\lambda, \lambda')|\rho(\lambda')/K_1^*\rho(\lambda),$$

where

$$\rho(\lambda) = \lim_{n\to\infty} \rho_n(\lambda),$$
$$\rho_n = \left[h_1(2\psi_1 - h_1) + (K_1^*\rho_{n-1})^2\right]^{1/2}, \qquad \rho_0 = |\psi_1|.$$

The function $\rho(\lambda)$ is defined exactly by the formulas above (i.e. it exists and is unique), it is nonnegative, belongs to the space L^∞, and satisfies the equation

$$\rho^2 = h_1(2\psi_1 - h_1) + (K_1^*\rho)^2.$$

Moreover,

$$\chi_1^* = \rho^2, \quad \mathbf{M}\xi_1^2 = (|f|, \rho_1)^2.$$

Conditions $K(\lambda, \lambda') \geq 0$, $h_1(\lambda) \geq 0$ imply $\rho = \psi_1$. If, in addition, $f(\lambda) \geq 0$, then the random variable ξ_1 is degenerate:

$$\xi_1 = (\varphi, h_1) = (f, \psi_1).$$

Remark 1. Concerning item III of Lemma 4.2 see [17], [73], [74]. For the case of elements with alternating signs the result on minimization of the variance for collision estimator was first obtained in [17].

Remark 2. The collision estimator (4.17) is often called “the basic estimator” for computing functional (4.18) (see [22], [68], [70]).

Lemma 4.2 is employed to describe H-optimal collision estimator

$$\xi(x) = \sum_{n=0}^{N} Q_n h(x, \lambda_n) \tag{4.19}$$

which in turn is used to compute the family of functionals

$$J(x) = (\varphi, h(x)) = (f, \psi(x)),$$

where $\varphi = K\varphi + f$, $\psi(x) = K^*\psi(x) + h(x)$.

Suppose that $\mathcal{D}^m h(x) \in L^\infty$, $\mathcal{D}^m\psi(x) \in L^\infty$, $\mathcal{D}^m K^*\psi(x) = K^*\mathcal{D}^m\psi(x)$, $\mathcal{D}^m J(x) = (\varphi, \mathcal{D}^m h(x)) = (f, \mathcal{D}^m\psi(x))$ for $x \in X$, $|m| \leq p$ (recall that p is the maximal order of the derivatives for the space $H = H^p(X, \{\mu_m\})$).

Theorem 4.2. *Assume that* $\|K_1\| < 1$ *and* $<h, (2\psi - h)> \geq 0$. *Then the Markov chain for the H-optimal collision estimator (4.19) is defined by formulas*

$$\begin{aligned} \pi(\lambda) &= |f(\lambda)|\rho(\lambda)/(|f|, \rho), \\ p(\lambda, \lambda') &= |K(\lambda, \lambda')|\rho(\lambda')/K_1^*\rho(\lambda), \end{aligned} \tag{4.20}$$

where nonnegative function $\rho(\lambda)$ *can be found as the unique solution of the equation*

$$\rho = \left[< h, (2\psi - h) > +(K_1^*\rho)^2\right]^{1/2}.$$

Moreover, for the H-optimal collision estimator we have

$$(\mathbf{M}\|\xi\|_H)^2 = \mathbf{M}\|\xi\|_H^2 = (|f|, \rho)^2.$$

Proof of the theorem mainly follows the proof of the statement from [1, p. 162]. First of all, we can easily verify that

$$\mathbf{M}\|\xi\|_H^2 = (\chi, < h, (2\psi - h) >).$$

Then we prove that there exist $\rho \geq 0$ and h_1 (unique in L^∞) such that

$$h_1(2\rho - h_1) =< h, (2\psi - h) >, \tag{4.21}$$

$$\rho = K_1^* + h_1. \tag{4.22}$$

Formulas (4.21), (4.22) imply

$$\rho^2 =< h, (2\psi - h) > +(K_1^*\rho)^2. \tag{4.23}$$

Since under conditions $\|K_1\| < 1$, $g =< h, (2\psi - h) > \geq 0$ operator

$$G(\rho) = \left[(K_1^*\rho)^2 + g\right]^{1/2}$$

is contracting:

$$\begin{aligned} |G(\rho_1) - G(\rho_2)| &\leq \||K_1^*\rho_1| - |K_1^*\rho_2|\| \leq \\ &\leq |K_1^*\rho_1 - K_2^*\rho| \leq \|K_1^*\| \; \|\rho_1 - \rho_2\|, \end{aligned}$$

then nonnegative solution of (4.23) exists and is unique. Obviously,

$$< h, (2\psi - h) > \leq \rho^2.$$

Setting now

$$h_1 = \rho - K_1^* \rho$$

we get (4.21) and (4.22).

Thus, we have

$$\mathbf{M}\|\xi\|_H^2 = (\chi, h_1(2\rho - h_1)) = \mathbf{M}\xi_1^2,$$

where ξ_1 is the collision estimator for the functional

$$J_1 = (\bar{\psi}, h_1) = (|f|, \rho).$$

Finally, it is easy to verify that for the Markov chain (4.20) the collision estimator ξ_1 is degenerate (cf. [22] and Lemma 4.2),

$$\mathbf{M}\xi_1^2 = (\bar{\varphi}, h_1)^2 = (|f|, \rho)^2,$$

and, hence, it gives the minimum of H-deviation of functional estimator (4.19). Theorem 4.2 has been proved.

Remark. For $H = L^2$ this theorem was proved in [75] with the help of recurrent relations and Bellman principle.

Lemma 4.3. *Under conditions of Theorems 4.1 and 4.2, the H-deviation of the H-optimal collision estimator does not exceed the H-deviation of the H-optimal absorption estimator.*

Proof. It is sufficient to show that

$$\rho \leq \psi_0. \tag{4.24}$$

We have

$$\begin{aligned} G(\psi_0) &= [(\psi_0 - h_0)^2 + g]^{1/2} = \\ &= [\psi_0 + 2(< h, \psi > - h_0\psi_0)]^{1/2}. \end{aligned}$$

Since, evidently,

$$\|\psi\|_H \leq K_1^* \|\psi\|_H + \|h\|_H,$$

and $\psi_0 = K_1^* \psi_0 + h_0$, then

$$\|\psi\|_H \leq \psi_0. \tag{4.25}$$

(see, for example, Lemma 1.1 from [74]). From (4.25) it follows that

$$< h, \psi > \leq h_0 \psi_0$$

and, hence, $G(\psi_0) \leq \psi_0$. By principle of contracting mapping $G_n(\psi_0) \to \rho$, and inequality (4.24) is fulfilled as the operator G is monotonic (for nonnegative functions). Lemma 4.3 has been proved.

Remark 1. Another way to prove Lemma 4.3 is to establish inequalities

$$\|\xi\|_H \leq \mathbf{M}\xi_0, \quad \mathbf{M}\|\xi\|_H^2 \leq \mathbf{M}\xi_0^2,$$

directly (here ξ is the collision estimator (4.19), ξ_0 is the collision estimator for the functional $(\bar{\varphi}, h_0) = (|f|, \psi_0)$ constructed on the same Markov chain as ξ) and then to make use of the Theorem 4.1.

Remark 2. In order to compute the family of the functionals $J(x) = (\varphi, h(x))$, where φ is a solution of the nonlinear integral equation

$$\varphi(\lambda) = \int_\Lambda \dots \int_\Lambda K(\lambda, \gamma_1 \dots \gamma_m) \prod_{j=1}^{m} \varphi(\gamma_j) d\gamma_j + f(\lambda),$$

we can use the analogues of the collision and absorption estimators that have just been considered (see [17], [22], [74], [75]). In this case the H-optimal estimators can be constructed in the same way as for linear integral equations of the second kind, and it should not involve appreciable difficulties.

4.3. Convergence of collision and absorption functional estimators

Analysis of convergence of collision and absorption estimators for computing a family of functionals

$$J(x) = (\varphi, h(x)) = (f, \varphi(x)),$$
$$\varphi = K\varphi + f, \quad \psi(x) = K^*\psi(x) + h(x),$$

can be performed on the basis of Statement 4.1. We present, as an example, a convergence criterion for the space of continuous functions $C(X)$, where X is a compact set in $\mathbb{R}^k$. As above, we denote by K_1 the integral operator with the kernel

$$K_1(\lambda', \lambda) = |K(\lambda', \lambda)|.$$

Statement 4.2. *Let $\xi(x)$ be an unbiased collision (absorption) estimator for $J(x)$ with finite variance: $\mathbf{M}\xi^2(x) < \infty, x \in X$. Suppose that there exist a number $\alpha > 0$ and a function $\bar{h}(\lambda) \geq 0$, $\bar{h}(\lambda) \in L^\infty$, such that*

$$|h(x,\lambda) - h(y,\lambda)| \leq \bar{h}(\lambda)|x-y|^\alpha, \quad x, y \in X, \tag{4.26}$$

and the collision (absorption) estimator for the quantity

$$(\varphi_1, \bar{h}) = (|f|, \psi_1),$$
$$\varphi = K_1\varphi_1 + |f|, \quad \psi_1 = K_1^*\psi_1 + \bar{h},$$

has a finite variance for the same Markov chain as for the estimator $\xi(x)$. Then the deviation in $C(X)$-norm of the functional estimate

$$J_m^*(x) = m^{-1}\sum_{j=1}^{m} \xi_j(x),$$

from $J(x)$ is of the order $m^{-1/2}$ in probability:

$$\mathbf{P}\left\{\|J_m^* - J\|_{C(X)} \leq am^{-1/2}\right\} \underset{m\to\infty}{\to} \text{const}(a).$$

(Here $\xi_j(x)$ are independent realizations of $\xi(x)$.)

Proof. It is sufficient to establish that the conditions in item (b) of Statement 4.1 are satisfied. It is obvious that

$$\mathbf{M}\varepsilon^2(x) = \mathbf{M}\xi^2(x) - J^2(x) < \infty,$$

for $\varepsilon(x) = \xi(x) - J(x)$. Inequality (4.26) implies

$$|J(x) - J(y)| \leq c|x-y|^\alpha, x, y \in \Lambda,$$

Indeed,

$$|J(x) - J(y)| \leq \int\limits_X \varphi(\lambda)\,[h(x,\lambda) - h(y,\lambda)]\,d\lambda,$$

$$\leq \left[\int\limits_X \varphi(\lambda)\bar{h}(\lambda)d\lambda\right] |x-y|^\alpha.$$

It remains to make sure that

$$|\varepsilon(x) - \varepsilon(y)| \leq G(\omega)|x-y|^\alpha \quad \text{and} \quad \mathbf{M}G^2(\omega) < \infty.$$

We have

$$|\varepsilon(x) - \varepsilon(y)| \leq |\xi(x) - \xi(y)| + |J(x) - J(y)| \leq$$
$$\leq \bar{\xi}|x-y|^a + c|x-y|^\alpha = (\bar{\xi} + c)|x-y|^\alpha.$$

where $\bar{\xi}$ is the collision (absorption) estimator for the quantity

$$(\varphi_1, \bar{h}), \quad \varphi_1 = K_1\varphi_1 + |f|,$$

and the estimator is constructed on the same Markov chain as the estimator $\xi(x)$, namely

$$\bar{\xi} = \sum_{n=0}^{N} |Q_n|\bar{h}(\lambda_n)$$

for the collision estimator, and

$$\bar{\xi} = |Q_N|\bar{h}(\lambda_N)/g(\lambda_N)$$

for absorption estimator. The Statement 4.2 has been proved, as we have shown that

$$\mathbf{M}(\bar{\xi} + c)^2 < \infty.$$

4.4. On convergence and optimization of local estimators

In order to compute the solution $\varphi(\lambda)$ of the integral equation of the second kind

$$\begin{aligned}&\varphi = K\varphi + f, \quad \varphi, f \in L^1 = L^1(\Lambda),\\&K : L^1 \to L^1,\\&\varphi(\lambda) = \int_{\Lambda} K(\lambda', \lambda)\varphi(\lambda')d\lambda' + f(\lambda),\end{aligned} \tag{4.27}$$

one can use so-called *local estimates* (see, for example, [22], [74])

$$\varphi_m^*(x) = m^{-1}\sum_{j=1}^{m} \xi_j(x) \to \varphi(x), \quad x \in \Lambda, \tag{4.28}$$

where $\xi_j(x)$ are independent realizations of the random function $\xi(x)$ that is constructed on the Markov chain $\{\lambda_n\}_{n=0,1,2...}$:

$$\begin{aligned}&\text{a)}\quad \xi(x) = Q_N K(\lambda_N, x)/g(\lambda_N) + f(x),\\&\text{b)}\quad \xi(x) = \sum_{n=0}^{N} Q_n K(\lambda_n, x) + f(x).\end{aligned}$$

The version a) corresponds to the absorption estimate, and version b) to the collision estimate.

It is not difficult to derive the unbiasedness conditions for these estimators and the corresponding expressions for variances

$$\begin{aligned}\text{a)}\quad \mathbf{D}\xi(x) &= (\chi, h_x^2/g) - (\varphi, h_x) =\\&= (f^2/\pi, \chi_x^*) - (f, \psi_x)^2,\\\text{b)}\quad \mathbf{D}\xi(x) &= (\chi, 2h_x\psi_x - h_x^2)^2 - (\varphi, h_x)^2 =\\&= (f^2/\pi, \theta_x^*) - (f, \psi_x)^2,\end{aligned}$$

where

$$\begin{aligned}&h_x(.) = K(., x), \qquad \psi_x = K^*\psi_x + h_x,\\&\psi_x(\lambda) = \int_L K(\lambda, \lambda')\psi_x(\lambda')d\lambda' + K(\lambda, x),\\&\chi = L\chi + f^2/\pi, \quad \chi_x^* = L^*\chi_x^* + h_x^2/g,\\&\theta_x^* = L^*\Theta_x^* + h_x(2\psi_x - h_x),\\&L\chi(\lambda) = \int \left[K^2(\lambda', \lambda)/p(\lambda', \lambda)\right]\chi(\lambda')d\lambda'.\end{aligned}$$

Theorems 4.1 and 4.2 enable us to obtain the H-optimal local estimators. For example, the Markov chain which realizes the H-optimal absorption estimator is defined by formulas

$$\pi(\lambda) = |f(\lambda)|\psi_0(\lambda)/(|f|, \psi_0),$$
$$p(\lambda', \lambda) = |K(\lambda', \lambda)|\psi_0(\lambda)/\psi_0(\lambda'),$$
$$g(\lambda) = h_0(\lambda)/\psi_0(\lambda),$$

where $\psi_0 = K_1^*\psi_0 + h_0$, $h_0(\lambda) = \|K(\lambda, .)\|_H$.

Convergence of local estimates can be studied by using Statement 4.1. Let us present the corresponding result for the space $C(\Lambda)$, Λ is a compactum in $\mathbb{R}^k$ (compare with Theorem 3 from [121]).

Statement 4.3. *Let $\xi(x)$ be an unbiased collision (absorption) local estimator for the solution $\varphi(x)$ of the integral equation of the second type (4.27) and $\mathbf{M}\xi^2(x) < \infty$, $x \in \Lambda$. Suppose that there exist numbers $\alpha > 0$, $c > 0$ and a function $\bar{h}(\lambda) \geq 0$, $\bar{h}(\lambda) \in L^\infty$, such that*

$$1) \quad |f(x) - f(y)| \leq c|x - y|^\alpha, \quad x, y \in \Lambda,$$
$$2) \quad |K(\lambda, x) - K(\lambda, y)| \leq \bar{h}(\lambda)|x - y|^\alpha, \quad x, y \in \Lambda,$$

and the collision (absorption) local estimator for the quantity

$$(\varphi_1, \bar{h}) = (|f|, \psi_1),$$
$$\varphi_1 = K_1\varphi_1 + |f|, \quad \psi_1 = K_1^*\psi_1 + \bar{h},$$

constructed on the same Markov chain as the estimatior $\xi(x)$ has a finite variance. (As usual, by K_1 we denote here the integral operator with the kernel $K_1(\lambda', \lambda) = |K(\lambda', \lambda)|$.) Then the functional estimate $\varphi_m^*(x)$ (cf. (4.28)) has a deviation in $C(\Lambda)$-norm from $\varphi(x)$ of the order $m^{-1/2}$ in probability.

Statement 4.3 can be proved similar to Statement 4.2.

4.5. Additional remarks

4.5.1. On convergence in the method of dependent tests

Let us focus again our attention on convergence of estimates

$$J_n^*(x) = \sum_{i=1}^{n} \xi_i(x)/n,$$

(cf. Item. 4.1). According to the law of large numbers, $J_n^*(x)$ converges to $J(x) = \mathbf{M}\xi(x)$ with probability one as $n \to \infty$. We consider the random function

$$w_n(x) = n^{-1/2} \sum_{i=1}^{n} \varepsilon_i(x),$$

where $\varepsilon_i(x)$ are independent realizations of the function

$$\varepsilon(x) = \xi(x) - J(x).$$

According to the central limit theorem, distributions of the vectors $(w_n(x_1), \dots, w_n(x_m))$ weakly converge to the Gaussian distributions with mean zero and covariances

$$K_{ij} = \mathbf{M}\varepsilon(x_i)\varepsilon(x_j), \quad i, j = 1, \dots, m,$$

if and only if $\mathbf{M}\varepsilon^2(x_i) < \infty$, $i = 1, \dots, m$.

However, the weak convergence in a functional space F of $w_n(x)$ to a Gaussian random function is of the most interest because in this case we can state that the deviation of J_n^* from J in the norm of this functional space is of the order $n^{-1/2}$ in probability:

$$\mathbf{P}\Big\{\|J_n^* - J\|_F c n^{-1/2}\Big\} \to \mathbf{P}\Big\{\|w\| \leq c\Big\}, \quad n \to \infty,$$

where w is the limit Gaussian function with mean zero and correlation function

$$\mathbf{M}w(x)w(y) = \mathbf{M}\varepsilon(x)\varepsilon(y).$$

If a measure in the functional space F is uniquely defined by its finite-dimensional distributions, then the weak compactness of w_n along with convergence of finite-dimensional distributions ensures the weak convergence of w_n to the Gaussian function w. In addition, the uniform integrability of the random variables

$$n\|J_n^* - J\|_F^2 = n^{-1} \left\| \sum_{i=1}^{n} \varepsilon_i \right\|_F^2$$

implies the convergence of the moments:

$$n\mathbf{M}\|J_n^* - J\|_F^2 \to \mathbf{M}\|w\|_F^2,$$

and the value $\mathbf{M}\|w\|_F^2$, which is called in Section 4.1 the F-deviation of estimator ξ, can be used as a measure of the estimator efficiency.

The conditions for the functional convergence in Statement 4.1 allow several generalizations that are of prime interest for Monte Carlo methods.

In order to investigate the convergence of vector algorithms which are actively developed in the theory of Monte Carlo methods (see,. e.g. [74]), of interest are conditions for weak convergence when space F is a space of vector-valued functions, i.e. when F can be represented as the product of spaces: $F = F_1 \times \ldots \times F_s$. Statement 4.1 can be easily generalized by using familiar arguments (see [13] and [60]). Conditions for weak convergence of vector-valued functional estimates are actually reduced to those of Statement 4.1 for each component of the vector function ε.

Other generalizations of conditions for weak convergence of functional Monte Carlo estimates, which are important for applications, can be obtained by using Borovkov's results [14]. The approach proposed in this paper enables us to obtain, for example, the conditions for functional convergence of $n^{1/2} \sum_{i=1}^{n} \varepsilon_i(x)$ to a continuous Gaussian process when the realizations of $\varepsilon(x)$ are not continuous.

Determination of the error of the estimates J_n^* is reduced to analysis of the rate of convergence in the central limit theorem. This problem has been studied in detail for the scalar (non-functional) case (see, e.g. [37], [101], [127]). In particular, if $\mathbf{M}|\varepsilon|^3 < \infty$, then

$$|\mathbf{P}([J_n^* - J] < n^{-1/2}x) - \Phi(x)| \leq A\frac{\mathbf{M}|\varepsilon|^3}{(\mathbf{M}|\varepsilon|^2)^{3/2}n^{1/2}(1 + |x|^3)}$$

where the constant A does not exceed 0.7655 (see [37]). Thus,

$$P(|J_n^* - J| < y) \in \left[\operatorname{erf}(n^{1/2}y) - \delta_n, \operatorname{erf}(n^{1/2}y) + \delta_n\right], \quad y > 0,$$

where

$$\delta_n = 2A\frac{\mathbf{M}|\varepsilon|^3}{(\mathbf{M}|\varepsilon|^2)^{3/2}n^{1/2}(1 + n^{3/2}|y|^3)},$$

$$\operatorname{erf}(x) = 2[\Phi(x) - 0.5],$$

and Φ is the standard normal distribution function.

For a functional case we would like to have some analytical tools to estimate the quantity

$$|\mathbf{P}(\|J_n^* - J\|_F < y) - \mathbf{P}(\|w\|_F < n^{1/2}y)|$$

where w is a Gaussian process with zero mean and corresponding correlation function. Moreover, here one faces the problem of calculating the probability $\mathbf{P}(\|w\|_F < c)$.

For the functional spaces F different from Sobolev's Hilbert spaces the rate of convergence

$$n\mathbf{M}\|J_n^* - J\|_F^2 \to \mathbf{M}\|w\|_F^2, \quad \text{as } n \to \infty.$$

is of interest as well.

Thus, the problem of analysis of convergence rate in the method of dependent tests is a topical one and calls for further investigation.

4.5.2. Discrete and generalized versions of optimization for Monte Carlo estimators in Sobolev's Hilbert spaces

We consider the segment $[a, b]$ as a set of the parameters X. Assume that functions on X from the space H^p are approximated by vectors of N components and the scalar product in H^p is approximated by its discrete analogue:

$$< f, g >= \sum_{m=0}^{p} \sum_{j=1}^{N-m} (D_m f)_j (D_m g)_j \mu_m^j, \tag{4.29}$$

where $\mu_m^j \geq 0$; f and g are vectors from $\mathbb{R}^N$; $(D_m f)_j$ is the j-th component of vector $D_m f$, and $N \times N$-matrices D_m are discrete analogues of operators of m-order differentiation.

Exercise 4.2. Find conditions for which (4.29) satisfies the axioms of scalar product. What variants can you propose to define matrices D_m as discrete analogues of differentiation operators?

We denote by $(.,.)$ the Euclidean scalar product in $\mathbb{R}^N$:

$$(f,g) = \sum_{j=1}^{N} f_j g_j.$$

Then we can write

$$< f,g >= \sum_{m=0}^{p} (L_m f, L_m g),$$

where $(L_m f)_j = (\mu_m^j)^{-1/2}(D_m f)_j$.

Thus, we consider $\mathbb{R}^N$ with the scalar product (4.29) as a space F. For F-deviation of Monte Carlo estimator $\xi = (\xi_1, \ldots, \xi_N)$ we have

$$\begin{aligned}
\mathrm{Dev}(\xi, F) &= \mathbf{M} < \xi - \mathbf{M}\xi, \xi - \mathbf{M}\xi >= \\
&= \mathbf{M} \sum_{m=0}^{p} (L_m[\xi - \mathbf{M}\xi], L_m[\xi - \mathbf{M}\xi]) = \\
&= \mathbf{M} \sum_{m=0}^{p} ([\xi - \mathbf{M}\xi], L_m^* L_m[\xi - \mathbf{M}\xi]) = \\
&= \mathbf{M}\,([\xi - \mathbf{M}\xi], A[\xi - \mathbf{M}\xi]) = \mathrm{tr}(AR),
\end{aligned}$$

where $A = \sum_{m=0}^{p} L_m^* L_m$ is a positive definite matrix, and R is the correlation matrix of the vector ξ,

$$R = \mathbf{M}[\xi - \mathbf{M}\xi][\xi - \mathbf{M}\xi]^T.$$

Generally, the criterion that minimizes the value of $\mathrm{tr}(AR)$, where $A > 0$ and R is a correlation matrix of the vector ξ, is called the criterion of linear optimality (see [23]). The equality $\mathrm{Dev}(\xi, F) = \mathrm{tr}(AR)$ implies that the criterion of F-optimality in the above case coincides with the criterion of linear optimality with the matrix A of a special form.

On the other hand, a criterion of linear optimality is a criterion of F-optimality in the space $F = \mathbb{R}^N$ with the norm $\|f\|_A = (f, Af)$ because

$$\mathrm{Dev}(\xi, F) = \mathbf{M}\|\xi - \mathbf{M}\xi\|_A^2 = \mathbf{M}\,(\xi - \mathbf{M}\xi, A[\xi - \mathbf{M}\xi]) = \mathrm{tr}(AR).$$

The assertions about the optimal Monte Carlo estimators in Sobolev's Hilbert spaces (Lemmas 4.1, 4.3 and Theorems 4.1, 4.2) remain valid for $F = \mathbb{R}^N$ with the scalar product $< f,g >= (f, Af)$.

Moreover, another Hilbert space can be considered as functional space H. The chain of equalities (4.11) is fundamental here. In particular, we may take as functional space H a subspace in $L^2(X,\mu)$ such that $A(H) \subset H$ holds for a positive operator $A : L^2(X,\mu) \to L^2(X,\mu)$. Then for the space H with the scalar product

$$< f, g >= \int_x f(x) A g(x) \mu(dx),$$

the assertions of Lemmas 4.1, 4.3 and Theorems 4.1, 4.2 remain valid if the conditions (A)-(C) in Item 4.2.1 are appropriately modified. These assertions are also valid for spaces with the scalar product of the form

$$< f, g >= \sum_m \int_x L_m f(x) L_m g(x) \mu_m(dx),$$

where $L_m : L^2(X,\mu_m) \to L^2(X,\mu_m)$ are some linear operators.

Remark. One can assume that results obtained by G.A. Mikhailov [74] concerning minimax Monte Carlo estimators are naturally generalized for the problem of minimizing the quantity

$$\max_{|m|\leq p,\, x\in X} \mathbf{D}\mathcal{D}^m \xi(x). \tag{4.30}$$

If there exist the "worst" distributions of μ_m which equalize the values $\mathbf{D}\mathcal{D}^m\xi(x)$ for the estimator ξ which is $H^p(X,\{\mu_m\})$-optimal, then the estimator ξ minimizes quantity (4.30).

Task 4.1. Study the hypothesis formulated in the latter Remark.

Task 4.2. Analyse the multi-level method of dependent tests proposed by S. Heinrich [40], [41], [42]. Study the problem of convergence and optimization of this method for different functional spaces.

Task 4.3. Find representations for different norms of a Gaussian vector (process) in terms of its expectation and covariances (see Remark 3 in Section 4.1)

Task 4.4. Analyse the rate of convergence of the expressions

$$|\mathbf{P}(\|J_n^* - J\|_F < y) - \mathbf{P}(\|w\|_F < n^{1/2}y)|,$$

$$n\mathbf{M}\|J_n^* - J\|_F^2 - \mathbf{M}\|w\|_F^2$$

to zero as $n \to \infty$ for various functional spaces (see Item. 4.5.1).

Task 4.5. Formulate and prove the generalizations of assertions presented in this Chapter for vector-valued Monte Carlo estimators.

Task 4.6. Obtain the conditions of weak convergence for Monte Carlo estimates on the basis of results from [14].

Task 4.7. Formulate and prove the extensions of Statements 4.2 and 4.3 for the spaces of continuously differentiable functions.

Task 4.8. Study convergence of unbiased functional Monte Carlo estimates for spaces which we have not considered in the Chapter (e.g., spaces without discontinuities of the second kind or non-Hilbert Sobolev's spaces; in particular, see [12], [35], [63]).

Task 4.9. Study the problem of minimizing relative error for the functional Monte Carlo estimates (see remark in Item 4.2.2). Consider different approaches to define relative error of functional estimates.

Appendix A

Gaussian distributions: properties and simulation

A.1. Gaussian distributions

A random vector $\xi = (\xi_1, \ldots, \xi_n)^T$ is called Gaussian if its characteristic function can be written in the form

$$\mathbf{M}e^{i<\lambda,\xi>} = \exp\left(i < \lambda, m > - < \lambda, R\lambda > /2\right), \quad \lambda \in \mathbb{R}^n,$$

where $m = \mathbf{M}\xi = (\mathbf{M}\xi_1, \ldots, \mathbf{M}\xi_n)^T$ is a vector of mean values,

$$R = \mathbf{M}(\xi - m)(\xi - m)^T = [\mathbf{M}(\xi_i - m_i)(\xi_j - m_j)]$$

is a correlation matrix and $< .,. >$ denotes the scalar product in $\mathbb{R}^n$. When the matrix R is nonsingular, the Gaussian distribution has probability density in $\mathbb{R}^n$:

$$f(x) = [\det(2\pi R)]^{-1/2} \exp\left[- < (x - m), R^{-1}(x - m) > /2\right], \quad x \in \mathbb{R}^n.$$

The Gaussian distributions satisfy several remarkable properties: non-correlatedness implies independence; a linear transformation of a Gaussian vector has a Gaussian distribution, too; the optimal mean square approximation of a Gaussian variable by a Gaussian family is a linear approximation.

For simulation of a pair of independent standard normal (i. e., Gaussian with zero mean and unit variance) random variables ε_1, ε_2 one can use the formulas

$$\varepsilon_1 = (-2\ln\alpha)^{1/2}\cos(2\pi\beta), \quad \varepsilon_2 = (-2\ln\alpha)^{1/2}\sin(2\pi\beta),$$

where α, β are independent random variables uniformly distributed in the interval $(0, 1)$. A detailed information on simulation of random variables can be found in [19], [22], [34], [39], [105], [111].

If $\xi = A\varepsilon$ is a linear transformation of a zero-mean Gaussian vector ε with correlation matrix R_ε, then the vector ξ is a Gaussian one with mean zero and correlation matrix $R = AR_\varepsilon A^*$. This fact is used for simulation of Gaussian vectors: let ε be a vector of independent standard normal variables ($R_\varepsilon = I$ is the identity matrix), then if we know a representation of R in the form AA^*, we can simulate the vector ξ by the formula $A\varepsilon$.

If the matrix A is self-conjugate, $A = A^*$, then $A = R^{1/2}$, and the matrix A can be approximately computed by the recurrent procedure

$$A_0 = 0, \quad A_{n+1} = A_n + \left(2\|R\|^{1/2}\right)^{-1} (R - A_n^2)$$

(see, e.g., [104]). Another way to find A is to use the representation $A = W(D)^{1/2}W^*$, where D is a diagonal matrix with eigenvalues of the matrix R as diagonal elements, and W is a matrix, whose columns are eigenvectors of the matrix R, $R = WDW^*$.

If we require that the matrix A be a lower triangular one, then it gives us the well known recurrent formulas for the matrix elements A_{rt} (see, for example, [22]):

$$\begin{aligned}
A_{11} &= (R_{11})^{1/2}, \\
A_{rt} &= \left(R_{rt} - \sum_{k=1}^{t-1} A_{rk}A_{tk}\right) / A_{tt}, \quad t = 1, \ldots, r-1, \\
A_{rr} &= \left(R_{rr} - \sum_{t=1}^{r-1} A_{rt}^2\right)^{1/2}, \quad r = 2, \ldots, s.
\end{aligned}$$

Exercise A.1. Prove that decomposition of a positive definite matrix $R = AA^*$ is uniquely defined accurate to a unitary operator: $R = AA^* = AU(AU)^*$, $UU^* = I$.

Task A.1. a) Offer your own method of decomposition $R = AA^*$, different from the above methods.

b) Realize a few algorithms for the simulation of a Gaussian vector and compare their computer costs.

A.2. Conditional Gaussian distributions

Assume that a Gaussian vector $\xi \in \mathbb{R}^n$ with expectation m and correlation matrix R consists of two components:

$$\xi = (\xi_1, \xi_2)^T, \quad \xi_1 \in \mathbb{R}^j, \quad \xi_2 \in \mathbb{R}^k, \quad j + k = n.$$

Then we can represent the mean vector and the correlation matrix in the following way:

$$m = (m_1, m_2)^T, \quad m_1 \in \mathbb{R}^j, \quad m_2 \in \mathbb{R}^k,$$

$$R = \begin{bmatrix} R_{11} & R_{12} \\ R_{21} & R_{22} \end{bmatrix},$$

where m_i are expectations and R_{ii} are correlation matrices of the vectors ξ_i, $i = 1, 2$.

Distribution of the vector ξ_1 provided that $\xi_2 = x$ is a Gaussian one with expectation

$$m(x) = m_1 + R_{12} R_{22}^{-1} (x - m_2)$$

and correlation matrix

$$R_{1|2} = R_{11} - R_{12} R_{22}^{-1} R_{21}.$$

(see, for example, [37]).

If the matrix R_{22} is singular and

$$\xi_2 = x \in m_2 + R_{22}(\mathbb{R}^k),$$

then the formulas for conditional expectation and correlation matrix take the form [1]

$$m(x) = m_1 + R_{12} R_{22}^{+} (x - m_2),$$
$$R_{1|2} = R_{11} - R_{12} R_{22}^{+} R_{21},$$

where R_{22}^{+} is the pseudoinverse to R_{22}. In case when x does not belong to support of distribution ξ_2, i. e., $x \notin m_2 + R_{22}(\mathbb{R}^k)$, the above formulas give the values of expectation and correlation matrix of the Gaussian distribution ξ_1 provided that $\|\xi_2 - x\|$ is minimal.

Task A.2. Construct and realize an algorithm for simulation of the Gaussian vector ξ_1 under condition that $\|\xi_2 - x\|$ is minimal. Use recurrent formulas from [1] for pseudoinversion of the correlation matrix R_{22}.

Exercise A.2. Derive formulas for simulation of a three-dimensional Gaussian vector (ξ_1, ξ_2, ξ_3) using conditional distribution of ξ_2 for the fixed ξ_1 and conditional distribution of ξ_3 for the fixed ξ_1, ξ_2. The formula for a two-dimensional inverse matrix will be recalled below:

$$\begin{bmatrix} a & b \\ c & d \end{bmatrix}^{-1} = \frac{1}{ad - cb} \begin{bmatrix} d & -b \\ -c & a \end{bmatrix}.$$

Exercise A.3. Derive formulas for simulation of values of the Wiener process $w(x)$ at the point $c \in (a, b)$ provided that the values of the Wiener process at points a, b are fixed. Compare the derived formulas with those from [22] for the case $c = (a + b)/2$. A hint: prove that conditional distribution of $w(c)$, provided that $w(a)=w_a$ and $w(b)=w_b$, is Gaussian with mean value

$$\frac{b - c}{b - a} w_a + \frac{c - a}{b - a} w_b$$

and variance

$$\frac{(b - c)(c - a)}{b - a}.$$

A.3. Schemes of moving average and autoregression

Moving average and autoregressive schemes are widely used to construct discrete models of stationary random processes and homogeneous fields. In this Section, we present in brief the basic relations for these models (for a particular study we refer to papers [2], [3], [15], [38], [66], [67] [69], [82], [103], [115]).

Complex-valued sequences of *moving average* are described by the formulas

$$x_n = \sum_{k=-\infty}^{\infty} A_k \varepsilon_{n-k} = \sum_{k=-\infty}^{\infty} A_{n-k} \varepsilon_k, \tag{A.1}$$

where ε_n is an orthogonal sequence of standard random variables ($\mathbf{M}\varepsilon_n = 0$, $\mathbf{M}|\varepsilon_n|^2 = 1$, $\mathbf{M}\varepsilon_n \bar{\varepsilon}_m = 0$ for $n \neq m$). The random **MA** sequence x_n is stationary, $\mathbf{M}x_n = 0$, and

$$R_n = \mathbf{M}x_{n+m}\bar{x}_m = \sum_{k=-\infty}^{\infty} A_{n+k}\bar{A}_k = \int_{-\pi}^{\pi} f(\lambda) e^{i\lambda n} d\lambda,$$

$$f(\lambda) = \frac{1}{2\pi} \sum_{n=-\infty}^{\infty} R_n e^{-i\lambda n} = \frac{1}{2\pi} \left| \sum_{k=-\infty}^{\infty} A_k e^{-i\lambda k} \right|^2,$$

$$a(\lambda) = \sum_{k=-\infty}^{\infty} A_k e^{-i\lambda k}, \quad |a(\lambda)|^2 = 2\pi f(\lambda).$$

Here R_n and $f(\lambda)$ are the correlation function and the spectral density of the sequence x_n, respectively; $a(\lambda)$ is the frequency characteristic of the linear transformation (A.1). For a specified spectral density we can build a **MA** model in the following way:

$$a(\lambda) = \sqrt{2\pi f(\lambda)}, \quad A_k = \frac{1}{2\pi} \int_{-\pi}^{\pi} a(\lambda) e^{i\lambda k} d\lambda.$$

Note, that this is the only method to construct **MA** model with a real-valued frequency characteristic $a(\lambda)$ (there exist a variety of models with complex-valued frequency characteristics). In case of real **MA** sequences for this model we have $a(\lambda) = a(-\lambda)$ (this follows from the equality $f(\lambda) = f(-\lambda)$ which is always fulfilled for real sequences), and the above formulas can be rewritten in the form:

$$R_n = \mathbf{M}x_{n+m}x_m = \sum_{k=-\infty}^{\infty} A_{n+k}A_k = \int_{-\pi}^{\pi} f(\lambda)\cos(\lambda n) d\lambda =$$

$$= 2\int_{0}^{\pi} f(\lambda)\cos(\lambda n) d\lambda,$$

$$f(\lambda) = \frac{1}{2\pi} \sum_{n=-\infty}^{\infty} R_n \cos(\lambda n) = \frac{1}{2\pi} \left[\sum_{k=-\infty}^{\infty} A_k \cos(\lambda k) \right]^2,$$

$$a(\lambda) = \sum_{k=-\infty}^{\infty} A_k \cos(\lambda k) = \sqrt{2\pi f(\lambda)},$$

$$A_k = \frac{1}{\pi}\int_0^{\pi} a(\lambda)\cos(\lambda k)d\lambda = \sqrt{\frac{2}{\pi}}\int_0^{\pi}\sqrt{f(\lambda)}\cos(\lambda k)d\lambda =$$

$$= \frac{1}{\pi}\int_0^{\pi}\left[\sum_{n=-\infty}^{\infty} R_n \cos(\lambda n)\right]^{1/2}\cos(\lambda k)d\lambda.$$

The moving average scheme is convenient for simulation of random fields (see [115]). Below we present the corresponding formulas for a random field on the plane:

$$x_{nm} = \sum_{k=-\infty}^{\infty}\sum_{j=-\infty}^{\infty} A_{kj}\varepsilon_{n-k,m-j} = \sum_{k=-\infty}^{\infty}\sum_{j=-\infty}^{\infty} \varepsilon_{kj}A_{n-k,m-j},$$

$$R_{nm} = \int_{-\pi}^{\pi}\int_{-\pi}^{\pi} f(\lambda,\nu)e^{i(\lambda n+\nu m)}d\lambda d\nu,$$

$$f(\lambda,\nu) = \frac{1}{(2\pi)^2}\sum_n\sum_m R_{nm}e^{-i(\lambda n+\nu m)} =$$

$$= \frac{1}{(2\pi)^2}\left|\sum_k\sum_j A_{kj}e^{-i(\lambda k+\nu j)}\right|^2,$$

$$a(\lambda,\nu) = \sum_k\sum_j A_{kj}e^{-i(\lambda k+\nu j)} = \sqrt{(2\pi)^2 f(\lambda,\nu)},$$

$$A_{kj} = \frac{1}{(2\pi)^2}\int_{-\pi}^{\pi}\int_{-\pi}^{\pi} a(\lambda,\nu)e^{i(\lambda k+\nu j)}d\lambda d\nu.$$

In a simple case when

$$f(\lambda,\nu) = f_1(\lambda)f_2(\nu), \quad R_{nm} = R_n^1 R_m^2,$$
$$a(\lambda,\nu) = a_1(\lambda)a_2(\nu), \quad A_{kj} = A_k^1 A_j^2,$$

the whole situation is reduced to the formulas of "one-dimensional type".

A real-valued random sequence of *autoregression* is defined by the relation

$$x_n = \sum_{k=1}^{m} b_k x_{n-k} + c\varepsilon_n, \quad n = m+1, m+2, \ldots,$$

where ε_n is an orthogonal sequence of standard random variables. Basically in applications, specialists use Gaussian **AR** models when standard random variables ε_n are normal and independent of each other and of x_k, $k < n$.

The stability condition (the condition of existence of a stationary solution): if all roots of the characteristic equation

$$\lambda^m = \sum_{k=1}^{m} b_k \lambda^{m-k}$$

are less than one in absolute value, then for any $x_1, x_2, \ldots, x_m$, the sequence x_n is asymptotically stationary, i. e., for sufficiently large n the sequence x_n can be considered as stationary with mean zero, correlation function $R_n = \mathbf{M}x_{n+k}x_k$ satisfying the *Yule–Walker equations*

$$R_n = \sum_{k=1}^{m} b_k R_{n-k}, \quad n = 1, 2, \ldots,$$

variance

$$R_0 = \mathbf{D}x_n = c^2 + \sum_{k=1}^{m} b_k R_k$$

and spectral density

$$f(\lambda) = \frac{c^2}{2\pi |Q(e^{-i\lambda})|^2}.$$

Here

$$Q(z) = 1 - b_1 z - b_2 z^2 - \ldots - b_m z^m$$

Another formulation of stability condition is the following: zero points of the polynomial $Q(z)$ lie out of the unit circle.

Autoregressive schemes can be used for simulation of random fields as well. For example, a homogeneous random field x_{ij} on a plane with correlation function $R_{ij} = \mathbf{M}x_{00}x_{ij} = \sigma^2 \rho^{i+j}$ can be simulated by the formulas

$$\begin{aligned}
x_{00} &= \sigma\varepsilon_{00}, \\
x_{i0} &= \rho x_{i-1,0} + \sigma(1-\rho^2)^{1/2}\varepsilon_{i0}, \quad i \geq 1, \\
x_{0j} &= \rho x_{0,j-1} + \sigma(1-\rho^2)^{1/2}\varepsilon_{0j}, \quad j \geq 1, \\
x_{ij} &= \rho x_{i-1,j} + \rho x_{i,j-1} - \rho^2 x_{i-1,j-1} + \sigma(1-\rho^2)\varepsilon_{ij}, \quad i,j \geq 1.
\end{aligned}$$

Exercise A.4. a) Construct an autoregressive model of a homogeneous random field x_{ijk} in the three-dimensional space with correlation function $R_{ijk} = \sigma^2\rho^{i+j+k}$.

b) Construct **AR** models of homogeneous fields in m-dimensional spaces with correlation function $R(i_1, i_2, \ldots, i_m) = \sigma^2\rho_1^{i_1}\rho_2^{i_2}\ldots\rho_m^{i_m}$.

c) Show that for the correlation function $R_{ij} = \rho_1^i\rho_2^j$, the corresponding **AR** scheme has the following form:

$$\begin{aligned}
x_{ij} &= \rho_1 x_{i-1,j} + \rho_2 x_{i,j-1} - \rho_1\rho_2 x_{i-1,j-1} + \\
&\quad + (1-\rho_1^2-\rho_2^2+\rho_1^2\rho_2^2)^{1/2}\varepsilon_{ij}.
\end{aligned}$$

Task A.3. Study stability conditions for **AR** models of random fields.

A.4. Generalized Wiener process

The *generalized Wiener process* (*fractional Brownian motion*) with parameter α, $\alpha \in (0, 2]$, is defined as a Gaussian process $w_\alpha(t)$ with mean zero and correlation function

$$K_\alpha(t,s) = Mw_\alpha(t)w_\alpha(s) = (|t|^\alpha + |s|^\alpha - |t-s|^\alpha)/2.$$

For $\alpha = 1$, we obtain the "ordinary" Wiener process:

$$K_1(t,s) = Mw_1(t)w_1(s) = (t+s-|t-s|)/2 = \min(t,s), \quad t,s \geq 0.$$

It is easy to see that $w_\alpha(0) = 0$ and $\mathbf{D}w_\alpha(t) = |t|^\alpha$. The fractal dimension of realizations of a generalized Wiener process is equal to $2-\alpha/2$ (see, for example, [24]).

The following representations can be proved for the generalized Wiener process:

$$
\begin{aligned}
K_\alpha(t,s) &= (|t|^\alpha + |s|^\alpha - |t-s|^\alpha)/2 = \\
&= \frac{A^2}{2\pi}\int_{-\infty}^{+\infty} \frac{1}{|\lambda|^{\alpha+1}}(e^{i\lambda t}-1)\,(e^{-i\lambda s}-1)\,d\lambda = \\
&= \frac{A^2}{\pi}\int_0^{+\infty} \frac{1}{\lambda^{\alpha+1}}[\cos(\lambda(t-s)) - \cos(\lambda t) - \cos(\lambda s) + 1]\,d\lambda, \\
w_\alpha(t) &= \frac{A}{\sqrt{2\pi}}\int_{-\infty}^{+\infty} \frac{1}{|\lambda|^{\frac{\alpha+1}{2}}}(e^{i\lambda t}-1)\,dz(\lambda) = \\
&= \frac{A}{\sqrt{\pi}}\{\int_0^{+\infty} \frac{1}{\lambda^{\frac{\alpha+1}{2}}}(\cos\lambda t - 1)\,d\xi(\lambda) - \int_0^{+\infty} \frac{1}{\lambda^{\frac{\alpha+1}{2}}}\sin\lambda t\,d\eta(\lambda)\}.
\end{aligned}
$$

Here $z(\lambda)$ is a standard complex-valued Wiener process and $\xi(\lambda)$, $\eta(\lambda)$ are standard real-valued Wiener processes:

$$
\begin{gathered}
\mathbf{M}z(\lambda) = \mathbf{M}\xi(\lambda) = \mathbf{M}\eta(\lambda) = 0, \\
\mathbf{M}|dz(\lambda)|^2 = \mathbf{M}[d\xi(\lambda)]^2 = \mathbf{M}[d\eta(\lambda)]^2 = d\lambda, \quad z(-\lambda) = \bar{z}(\lambda), \\
z(\lambda) = \frac{1}{\sqrt{2}}\xi(\lambda) + i\frac{1}{\sqrt{2}}\eta(\lambda).
\end{gathered}
$$

For $\alpha = 1$, the constant A is equal to 1, otherwise it can be found from

$$
A^2 = \left\{\frac{2}{\pi}\int_0^\infty \frac{1}{\lambda^{\alpha+1}}(1-\cos\lambda)d\lambda\right\}^{-1} = \left\{-\frac{2}{\pi}\Gamma(-\alpha)\cos\frac{\alpha\pi}{2}\right\}^{-1}.
$$

Task A.4. a) Derive the above representations taking into account the following relations

$$
\begin{gathered}
\frac{1}{2}\int_0^\infty x^{\mu-1}(1-\cos 2bx)dx = \int_0^\infty x^{\mu-1}\sin^2 bx\,dx = -\frac{\Gamma(\mu)}{2^{\mu+1}}|b|^{-\mu}\cos\frac{\mu\pi}{2}, \\
-2 < \operatorname{Re}\mu < 0.
\end{gathered}
$$

b) Construct "spectral" models to simulate generalized Wiener processes.

c) What other numerical methods can you propose to simulate a fractional Brownian motion?

Below we present some exercises concerning generalized Wiener processes.

Exercise A.5. Prove that the random processes $w_\alpha(Ct)$ and $C^{\alpha/2}w_\alpha(t)$, $C > 0$, have the same finite-dimensional distributions (this property is called *self-similarity* with the scaling parameter $\alpha/2$).

Exercise A.6. a) Prove that the random process $\delta(t) = w_\alpha(t+h) - w_\alpha(t)$ is stationary with mean zero, correlation function

$$R(\tau) = \mathbf{M}\delta(t+\tau)\delta(t) = \frac{|\tau+h|^\alpha + |\tau-h|^\alpha}{2} - |\tau|^\alpha$$

and spectral density

$$\frac{A^2}{\pi}\frac{1}{|\lambda|^{\alpha+1}}(1-\cos\lambda h), \quad \lambda \in (-\infty, +\infty).$$

b) Prove that the random sequence $\delta(n)$ has the following spectral density

$$\frac{A^2}{\pi}(1-\cos\lambda h)\sum_{k=-\infty}^{+\infty}\frac{1}{|\lambda+2k\pi|^{\alpha+1}}, \quad \lambda \in (-\pi, \pi).$$

Using the above representation, show that

$$2(1-\cos\lambda)\sum_{k=-\infty}^{+\infty}\frac{1}{|\lambda+2k\pi|^2}$$

is equal to one for all λ.

c) Construct numerical methods on the basis of moving average and autoregressive schemes to simulate the sequence $\delta(n)$ and analyse these methods.

Appendix B

Solution of boundary value problems for linear systems of stochastic differential equations

B.1. General relations

Let us consider a linear vector stochastic differential equation (SDE) of the Ito type

$$dx(t) = A(t)x(t)dt + \Sigma(t)dw(t), \tag{B.1}$$

where $x(t) = (x_1(t), \ldots, x_k(t))^T$ is a vector-valued random process, $w(t) = (w_1(t), \ldots, w_k(t))^T$ is a vector of independent standard Wiener processes, $A(t)$ and $\Sigma(t)$ are $k \times k$ matrices whose elements are piecewise continuous functions. It is a well-known fact (cf., for example, [110] [116]) that if the random process $x(t)$ satisfies SDE (B.1) in the interval $t \in (a, b)$, then for $a < s < t < b$ we have

$$x(t) = \Phi(t,s)x(s) + \int_s^t \Phi(t,\tau)\Sigma(\tau)dw(\tau). \tag{B.2}$$

Here $\Phi(t, s)$ is the principal matrix associated with $A(t)$. The principal matrix has the following properties:

1) $\Phi(t,s) = \Phi(t,\tau)\Phi(\tau,s)$ for $s \le \tau \le t$;
2) the matrix $\Phi(t, s)$ is nonsingular for any t and s from (a, b); in addition, $\Phi(s,t) = \Phi^{-1}(t,s)$ and $\Phi(\tau,\tau)$ is the identity matrix;
3) $\dfrac{\partial\Phi(t,s)}{\partial t} = A(t)\Phi(t,s), \quad t \ge s;$
4) $\dfrac{\partial\Phi(t,s)}{\partial s} = -\Phi(t,s)A(s), \quad s \le t;$
5) $\Phi(t,s) = W(t)W^{-1}(s)$, where $W(t)$ is the fundamental solution matrix for the equation $x' = Ax$,

$$\frac{d}{dt}W(t) = A(t)W(t).$$

For the matrix-valued covariance function

$$B(t,s) = \mathbf{M}x(t)x^T(s)$$

we have

$$B(t,s) = \Phi(t,s)B(s,s), \quad t \geq s.$$

Consider a random vector

$$y_s^t = \int_s^t \Phi(t,\tau)\Sigma(\tau)dw(\tau).$$

The vector y_s^t is a Gaussian one with mean zero and correlation matrix

$$R_s^t = \int_s^t \Phi(t,\tau)\Sigma(\tau)\Sigma^T(\tau)\Phi^T(t,\tau)d\tau$$

(it is called Gram's matrix), furthermore (cf. [116]),

$$\frac{\partial R_s^t}{\partial s} = A(t)R_s^t + R_s^t A^T(t) + \Sigma(t)\Sigma^T(t), \quad t > s.$$

Lemma B.1. *Assume that a vector-valued random process $x(t)$ satisfies SDE (B.1) on the segment $[\alpha,\beta]$ and the matrix R_α^β is nonsingular. Then the distribution of the vector $x(\gamma)$, $\gamma \in (\alpha,\beta)$, provided that $x(\alpha) = x_\alpha$, $x(\beta) = x_\beta$ (the vectors x_α, x_β are nonrandom), is Gaussian with expectation*

$$m = \left[\Phi(\gamma,\alpha) - R_\alpha^\gamma \Phi^T(\beta,\gamma)(R_\alpha^\beta)^{-1}\Phi(\beta,\alpha)\right] x_\alpha + \\ + R_\alpha^\gamma \Phi^T(\beta,\gamma)(R_\alpha^\beta)^{-1} x_\beta \tag{B.3}$$

and correlation matrix

$$\mathbf{M}\left(x(\gamma) - m\right)\left(x^T(\gamma) - m^T\right) = \\ = R_\alpha^\gamma - R_\alpha^\gamma \Phi^T(\beta,\gamma)(R_\alpha^\beta)^{-1}\Phi^T(\beta,\gamma)R_\alpha^\gamma. \tag{B.4}$$

Proof. We can find from (B.2) that

$$x(\gamma) = \Phi(\gamma, \alpha)x_\alpha + y_\alpha^\gamma, \tag{B.5}$$
$$x(\beta) = \Phi(\beta, \alpha)x_\alpha + y_\alpha^\beta.$$

The random vector $(y_\alpha^\gamma, y_\alpha^\beta)^T$ is Gaussian with mean zero and correlation matrix

$$\begin{bmatrix} R_\alpha^\gamma & R_\alpha^\gamma \Phi^T(\beta, \gamma) \\ \Phi^T(\beta, \gamma) R_\alpha^\gamma & R_\alpha^\beta \end{bmatrix}.$$

Hence (see Item A2), conditional distribution of the vector y_α^γ, provided that $y_\alpha^\beta = x_\beta - \Phi(\beta, \alpha)x_\alpha$, is Gaussian with expectation

$$R_\alpha^\gamma \Phi^T(\beta, \gamma)(R_\alpha^\beta)^{-1} [x_\beta - \Phi(\beta, \alpha)x_\alpha]$$

and correlation matrix (B.4). To complete the proof, it remains to note that (B.5) results in (B.3).

Remark 1. If the matrix R_α^β is singular and $x_\beta \in \Phi(\beta, \alpha)x(\alpha) + R_\alpha^\beta \mathbb{R}^k$, then formulas (B.3), (B.4) remain valid with $(R_\alpha^\beta)^+$ instead of $(R_\alpha^\beta)^{-1}$.

Remark 2. For the scalar case ($k = 1$)

$$dx(t) = a(t)x(t)dt + \sigma(t)dw(t)$$

we have

$$\Phi(t, s) = \exp\left(\int_s^t a(\tau)d\tau\right),$$

$$R_s^t = \int_s^t \exp\left(2\int_\tau^t a(y)dy\right)\sigma^2(\tau)d\tau.$$

B.2. Boundary value problems for time-invariant linear systems

For SDE (B.1) with constant coefficients (the matrices $A(t) = A$ and $\Sigma(t) = \Sigma$ do not depend on t) we have

$$\Phi(t, s) = \exp(A(t - s)).$$

For the solution $x(t)$ of the SDE

$$dx(t) = Ax(t)dt + \Sigma w(t), \tag{B.6}$$

we can write down (see (B.2))

$$x(\beta) = Bx(\alpha) + C\varepsilon,$$

where $\beta > \alpha$, $B = \exp(A(\beta - \alpha))$, the matrix C satisfies the relation

$$R_\alpha^\beta = CC^T = \int_0^{\beta-\alpha} e^{A\tau}\Sigma\Sigma^T e^{A^T\tau} d\tau,$$

while $\varepsilon = (\varepsilon_1, \dots, \varepsilon_k)^T$ is a vector of independent standard normal random variables (dependent on the Wiener process $w(t)$).

If the matrix A is **stable**, i.e., real parts of the roots λ of the characteristic equation $\det(A-\lambda I) = 0$ are strictly negative, then the following equality is valid

$$CC^TA^T + ACC^T = B\Sigma\Sigma^TB^T - \Sigma\Sigma^T.$$

The assertion presented below can be proved directly or as a corollary of Lemma B.1.

Lemma B.2. *Assume that the vector-valued random process $x(t)$ is a solution of SDE (B.6) with a stable matrix A. Then the distribution of the vector $x(\gamma)$, $\gamma \in (\alpha, \beta)$, provided that $x(\alpha) = x_\alpha$, $x(\beta) = x_\beta$, is the Gaussian one with expectation*

$$m^* = (B_1 - K_1B_2^TK^{-1}B)x_\alpha + K_1B_2^TK^{-1}x_\beta \tag{B.7}$$

and correlation matrix

$$R^* = \mathbf{M}\,(x(\gamma) - m^*)\,(x(\gamma) - m^*)^T = K_1 - K_1B_2^TK^{-1}B_2K_1, \tag{B.8}$$

where $B_1 = \exp((\gamma - \alpha)A)$, $B_2 = \exp((\beta - \gamma)A)$, $B = B_1B_2 = B_2B_1$,

$$K_1A^T + AK_1 = B_1\Sigma\Sigma^TB_1^T - \Sigma\Sigma^T, \tag{B.9}$$

$$K_2A^T + AK_2 = B_2\Sigma\Sigma^TB_2^T - \Sigma\Sigma^T, \tag{B.10}$$

$$KA^T + AK = B\Sigma\Sigma^TB^T - \Sigma\Sigma^T, \tag{B.11}$$

$$K = B_2K_1B_2^T + K_2.$$

(Here we assume that the matrix K is nonsingular. In particular, K is nonsingular if the matrix Σ is nonsingular.)

Remark 3. Solutions of equations (B.9)–(B.11) exist and they are unique, since the matrix

$$K(t) = \int_0^t e^{A\tau}\Sigma\Sigma^T e^{A^T\tau} d\tau =$$

$$= R_\alpha^{\alpha+t} = \int_\alpha^{\alpha+t} e^{A(\alpha+t-\tau)}\Sigma\Sigma^T e^{A(\alpha+t-\tau)} d\tau$$

is the unique solution of the Lyapunov equation (cf. [32])

$$K(t)A^T + AK(t) = e^{At}\Sigma\Sigma^T e^{A^T t} - \Sigma\Sigma^T.$$

For the numerical solution of the Lyapunov equation see, for example, [33], [43].

Remark 4. For the scalar SDE

$$dx(t) = axdt + \sigma dw(t), \quad a < 0, \tag{B.12}$$

formulas (B.7), (B.8) take the form

$$m^* = \frac{(b_2^2 - 1)b_1}{b^2 - 1}x(\alpha) + \frac{(b_1^2 - 1)b_2}{b^2 - 1}x(\beta),$$

$$R^* = \sigma^2 \frac{(b_2^2 - 1)(b_1^2 - 1)}{2a(b^2 - 1)},$$

$$b_1 = e^{(\gamma-\alpha)a}, \quad b_2 = e^{(\beta-\gamma)a}, \quad b = b_1 b_2 = e^{(\beta-\alpha)a},$$

and, hence, the value $x(\gamma)$ for the fixed $x(\alpha)$ and $x(\beta)$ can be simulated by the formula

$$x(\gamma) = m^* + \sqrt{R^*}\varepsilon,$$

where ε is a standard normal random variable (Figure B.1).

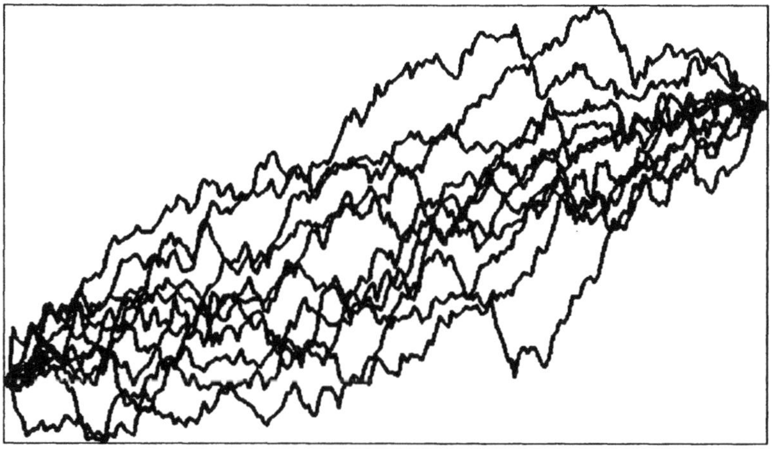

Figure B.1. A set of realizations of a random process that is the solution of the boundary value problem for the scalar linear SDE

For $\gamma = (\alpha + \beta)/2$, we have

$$m^* = \frac{b^{1/2}}{b+1}\left(x(\alpha) + x(\beta)\right),$$

$$R^* = \sigma^2 \frac{(b-1)}{2a(b+1)}.$$

Remark 5. For the Wiener process $w(t)$, the corresponding formulas for the conditional expectation and variance can be obtained from (B.12) as $a \to 0$, $\sigma \to 1$:

$$\mathbf{M}\left(w(\gamma) | w(\alpha) = w_\alpha, w(\beta) = w_\beta\right) = \frac{\beta - \gamma}{\beta - \alpha} w_\alpha + \frac{\gamma - \alpha}{\beta - \alpha} w_\beta,$$

$$\mathbf{D}\left(w(\gamma) | w(\alpha) = w_\alpha, w(\beta) = w_\beta\right) = \frac{(\beta - \gamma)(\gamma - \alpha)}{\beta - \alpha}.$$

Exercise B.1. To numerically solve the boundary value problem (i.e., to simulate a realization $x(t)$, $t \in [\alpha, \beta]$, integrally) we use the following property of the process $x(t)$: the conditional distribution of $x(t_0)$, $t_0 \in (t_1, t_2)$ for given values $x(t_1)$, $x(t_2)$ does not depend on the values $x(t)$ for $t < t_1$ and $t > t_2$. Prove this property.

B.3. On correctness of boundary value problems

Here we consider more general boundary conditions for the linear SDE and discuss their **correctness** (see Definition below).

Definition. A boundary value problem for the linear SDE

$$dx(t) = A(t)x(t)dt + \Sigma(t)dw(t), \quad t \in (\alpha, \beta), \tag{B.13}$$

with certain boundary conditions on the vectors $x(\alpha)$, $x(\beta)$ is said to be **correct** if the boundary conditions enable one to uniquely define all the finite-dimensional distributions of the process $x(t)$, $t \in [\alpha, \beta]$. The corresponding boundary conditions will be called **correct**, too.

Evidently, the boundary conditions are correct if and only if they enable one to uniquely define the joint distribution of the boundary values $x(\alpha)$, $x(\beta)$ satisfying the relation

$$x(\beta) = \Phi(\beta, \alpha)x(\alpha) + \xi,$$

where ξ is a normal random vector with mean zero and correlation matrix R_α^β. Thus, the numerical solution of a correct boundary value problem can be reduced to simulation of the boundary values $x(\alpha)$, $x(\beta)$. Then the conditional distributions of $x(t)$, $t \in (\alpha, \beta)$, are defined by Lemma B.1 (see, also, Remark 1 and Exercise B.1).

Task B.1. Consider the boundary conditions of the form

$$Ex(\alpha) + Hx(\beta) = \varphi, \tag{B.14}$$

where φ is n-dimensional vector and E, H are $n \times k$ matrices. Prove that the boundary value problem (B.13), (B.14) is correct if and only if the following two conditions are fulfilled:

1) $\text{rang}(D) = k$ (this implies $n \geq k$),

2) $P\varphi \in PHR_\alpha^\beta(\boldsymbol{R}^k)$,

where

$$P = I - DD^+, \quad D = E + H\Phi(\beta, \alpha).$$

Construct an algorithm to simulate the boundary values $x(\alpha)$, $x(\beta)$.

Task B.2. Prove the following assertion.

Lemma B.3. *Assume that $x \in \boldsymbol{R}^k$, A is $n \times k$ matrix, ξ is n-dimensional Gaussian random vector with expectation m and correlation matrix R.*

1. *On condition that there exists a solution of the linear system $Ax = \xi$, the distribution of the vector ξ is Gaussian with expectation*

$$Qm$$

and correlation matrix

$$QR,$$

where

$$Q = I - RP^T(PRP^T)^+P = I - R^{1/2}(PR^{1/2})^+P,$$

$P = I - AA^+ = Pr(\mathrm{Ker}(A^T))$ is the projector on the kernel of the matrix A^T. Moreover, the following condition

(a) $Pm \in PR(\boldsymbol{R}^n)$.

is necessary and sufficient for existence of vectors $y \in R(\boldsymbol{R}^n)$, from the support of distribution of the random vector ξ, for which the system $Ax = y$ is consistent.

2. *The equation $Ax = \xi$ uniquely defines distribution of the vector x if and only if (a) is fulfilled with the complementary condition*

(b) $\mathrm{rang}(A) = k$ *(and, hence, $n \geq k$).*

Moreover, in this case the vector $x = A^+\xi$ is Gaussian with expectation A^+Qm and correlation matrix $A^+QR(A^+)^T$.

To prove Lemma B.3, make use of Lemma 1.2 from Chapter 1 and the fact that the condition $\xi \in Im(A)$ means $(I - AA^+)\xi = 0$.

Using Lemma B.3, find the joint distribution of the vectors $x(\alpha)$, $x(\beta)$ for the correct boundary conditions (B.14).

Remark 6 (on correctness of boundary value problems for controllable systems). Assume that the matrix R_α^β is non-singular (in this case, the linear system $x'(t) = A(t)x(t) + \Sigma(t)u(t)$ is called controllable). Then it is easy to verify that the following boundary conditions are correct

$$1)\quad x(\alpha) = x_\alpha, \quad x(\beta) = x_\beta,$$

$$2)\quad x(\alpha) = x_\alpha, \quad R\,x(\beta) = r,$$

$$3)\quad L\,x(\alpha) = l, \quad x(\beta) = x_\beta.$$

Here x_α, x_β are arbitrary k-dimensional vectors and the linear systems $Rx = r$, $Lx = l$ are individually consistent.

The boundary conditions

$$L\,x(\alpha) = l, \quad R\,x(\beta) = r,$$

can be not correct, because sometimes they do not uniquely define the joint distribution of the vectors $x(\alpha)$, $x(\beta)$.

Remark 7 (on correctness of boundary value problems for the scalar time-invariant linear SDE). Let us consider a scalar autonomous linear SDE ($k = 1$)

$$dx(t) = A\,x(t)\,dt + \Sigma\,dw(t), \quad \alpha < t < \beta, \quad A \neq 0, \quad \Sigma \neq 0,$$

with the boundary conditions

$$Ex(\alpha) + Hx(\beta) = \varphi,$$

where E, H, φ are vectors of dimension $n \leq 2$. In this case, we have

$$\Phi(t,s) = e^{A(t-s)},$$

$$R_\alpha^\beta = \frac{\Sigma^2}{2A}[e^{2A(\beta-\alpha)} - 1].$$

Let us apply the necessary and sufficient conditions from Task B.1 for the correctness of the boundary value problem. Condition 1 can be written down in the form

$$D = E + He^{A(t-s)} \neq 0.$$

Condition 2 is always fulfilled for $n = 1$, since in this case we have $P = I - DD^+ = 0$. For $n = 2$, condition 2 can be formulated as follows: if the vectors E, H are linear dependent (proportional), then the vector φ must be proportional to the vector D (otherwise, there are no restrictions on the vector φ).

Remark 8 (on construction of realization of the Wiener process when solving the boundary value problem). In the case, when we are interested not only in the solution $x(t)$ of the boundary value problem (B.13), (B.14), but in the simulation of the Wiener process $w(t)$ from (B.13) as well, we ought to consider and to numerically solve the extended system

$$\begin{cases} dx(t) = A(t)\, x(t)\, dt + \Sigma(t)\, dw(t), \\ dy(t) = dw(t) \end{cases}$$

with the additional boundary condition

$$y(\alpha) = w(\alpha).$$

B.4. Stationary boundary value problems

Let us consider the time-invariant linear SDE

$$dx(t) = Ax(t)dt + \Sigma dw(t) \tag{B.15}$$

with a stable matrix A. It is a well-known fact that there exists a vector-valued stationary random process, which satisfies (B.15). The stationary solution $x(t)$ of equation (B.15) is a Gaussian process with mean zero and correlation function

$$R(\tau) = \mathbf{M}x(t+\tau)x^T(t) = \exp(A\tau)R_0,$$

where

$$R_0 = \int_0^\infty \exp(A\tau)\Sigma\Sigma^T \exp(A^T\tau)d\tau$$

is the unique solution of the Lyapunov equation

$$R_0A^T + AR_0 = -\Sigma\Sigma^T.$$

Now we consider the boundary value problem for equation (B.15) in the interval $t \in [\alpha, \beta]$ with boundary conditions (B.14) under an additional assumption that the process $x(t)$ is stationary (such a boundary value problem will be called **stationary**).

Task B.3. Prove, that the stationary boundary value problem (B.15) with boundary conditions (B.14) is correct if and only if

$$\tilde{P}\varphi \in \tilde{P}HR_{\alpha}^{\beta}(\boldsymbol{R}^k),$$

where

$$\tilde{P} = I - (DR_0)(DR_0)^+, \quad D = E + H\Phi(\beta, \alpha).$$

B.5. Boundary value problems for the second-order autonomous linear SDE

Let us study a scalar time-invariant linear SDE of the second order

$$\begin{aligned} &y''(t) + ay'(t) + by(t) = \sigma w'(t), \\ &a, b, \sigma > 0, \qquad \omega^2 = b - a^2/4 > 0, \end{aligned} \tag{B.16}$$

which corresponds to the vector equation (B.15) with the matrices

$$A = \begin{bmatrix} 0 & 1 \\ -b & -a \end{bmatrix}, \quad \Sigma = \begin{bmatrix} 0 & 0 \\ 0 & \sigma \end{bmatrix}.$$

In this case, we have

$$\exp\left\{\begin{bmatrix} 0 & 1 \\ -b & -a \end{bmatrix} t\right\} = \exp(-at/2) \times$$

$$\times \begin{bmatrix} \cos(\omega t) + \dfrac{a}{2\omega}\sin(\omega t) & \dfrac{1}{\omega}\sin(\omega t) \\ -\dfrac{b}{\omega}\sin(\omega t) & \cos(\omega t) - \dfrac{a}{2\omega}\sin(\omega t) \end{bmatrix},$$

$$\omega = \sqrt{b - a^2/4}.$$

and the matrix $K(t)$, satisfying the Lyapunov equation

$$\begin{aligned} K(t)A^T + AK(t) &= B(t)\Sigma\Sigma^T B^T(t) - \Sigma\Sigma^T, \\ B(t) &= \exp(At), \end{aligned}$$

has the following form

$$K(t) = \int_0^t B(\tau)\Sigma\Sigma^T B^T(\tau)d\tau = \begin{bmatrix} K_{11}(t) & K_{12}(t) \\ K_{21}(t) & K_{22}(t) \end{bmatrix},$$

$$K_{11}(t) = \frac{\sigma^2}{b}\left[\frac{1}{2a} - e^{-at}\frac{1}{8a\omega^2}\left(4b - a^2\cos(2\omega t) + 2a\omega\sin(2\omega t)\right)\right],$$

$$K_{22}(t) = \sigma^2\left[\frac{1}{2a} - e^{-at}\frac{1}{8a\omega^2}\left(4b - a^2\cos(2\omega t) - 2a\omega\sin(2\omega t)\right)\right],$$

$$K_{12}(t) = K_{21}(t) = \frac{\sigma^2}{4\omega^2}e^{-at}\left(1 - \cos(2\omega t)\right).$$

Obviously, the matrix $K(t)$ is non-singular for $t > 0$, otherwise the following "false" equality $y(t) = c_1 y'(t) + c_2$ is fulfilled. Therefore, the boundary conditions

$$\begin{aligned} y(\alpha) &= y_\alpha, \quad y(\beta) = y_\beta, \\ y'(\alpha) &= y'_\alpha, \quad y'(\beta) = y'_\beta \end{aligned} \tag{B.17}$$

are correct in the sense of the definition from Item B.3 (see Figure B.2).

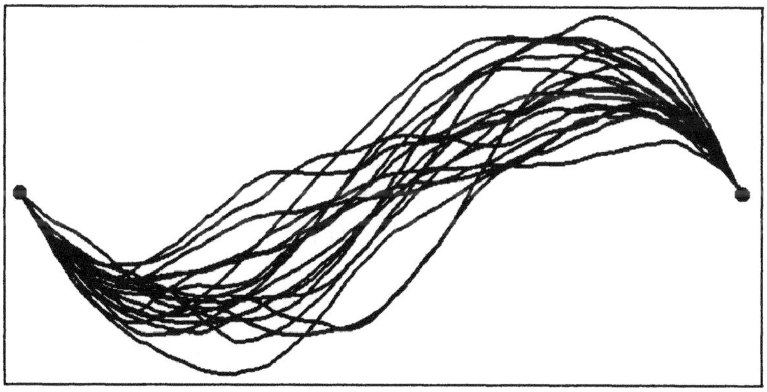

Figure B.2. A set of realizations of a random process that is the solution of the boundary value problem (B.16), (B.17) for the second-order autonomous linear SDE

Let us now consider the boundary conditions of the form

$$y(\alpha) = y_\alpha, \quad y(\beta) = y_\beta. \tag{B.18}$$

Task B.4. Prove, that the boundary value problem (B.16), (B.18) is not correct for

$$\omega(\beta - \alpha) = m\pi, \quad m = 1, 2, \dots . \tag{B.19}$$

The **stationary** boundary value problem (B.16), (B.18) is correct (even for condition (B.19)), because the stationarity of the process $y(t)$ enables one to find the conditional distribution of the vector $(y'(\alpha), y'(\beta))^T$ provided (B.18). It means that we can simulate the vector of boundary values and to reduce the problem to the correct boundary conditions (B.17) (see Figure B.3).

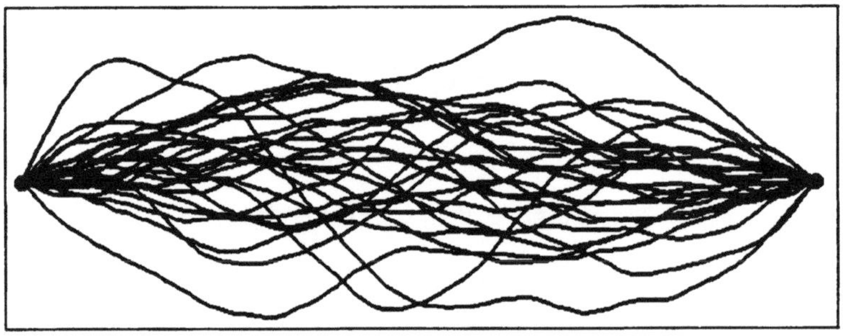

Figure B.3. A set of realizations of a random process that is the solution of the "stationary" boundary value problem (B.15), (B.18) for the second-order SDE

Additional explanation. For the stationary process

$$x(t) = \begin{bmatrix} y(t) \\ y'(t) \end{bmatrix},$$

which is the solution of SDE (B.16), we have

$$\mathbf{M}x(\beta)x^T(\alpha) = \exp\left(A(\beta - \alpha)\right) D = L,$$

where

$$D = \int_0^\infty e^{A\tau}\Sigma\Sigma^T e^{A^T\tau} d\tau = \lim_{t\to\infty} K(t) = \begin{bmatrix} \dfrac{\sigma^2}{2ab} & 0 \\ 0 & \dfrac{\sigma^2}{2a} \end{bmatrix}.$$

Therefore,

$$\mathbf{M}\begin{bmatrix} y'(\alpha) \\ y'(\beta) \\ y(\alpha) \\ y(\beta) \end{bmatrix} [y'(\alpha)y'(\beta)y(\alpha)y(\beta)] = \begin{bmatrix} D_{22} & L_{22} & D_{12} & L_{12} \\ L_{22} & D_{22} & L_{21} & D_{12} \\ D_{12} & L_{21} & D_{11} & L_{11} \\ L_{12} & D_{12} & L_{11} & D_{11} \end{bmatrix},$$

where L_{ij} and D_{ij} are elements of the matrices L and D, respectively.

The distribution of the vector

$$\begin{bmatrix} y'(\alpha) \\ y'(\beta) \end{bmatrix}$$

under condition (B.18) is Gaussian with expectation

$$R_{12}R_{22}^{-1}\begin{bmatrix} y(\alpha) \\ y(\beta) \end{bmatrix}$$

and correlation matrix

$$R_{11} - R_{12}R_{22}^{-1}R_{21},$$

where

$$R_{11} = \begin{bmatrix} D_{22} & L_{22} \\ L_{22} & D_{22} \end{bmatrix}, \quad R_{12} = \begin{bmatrix} D_{12} & L_{12} \\ L_{21} & D_{12} \end{bmatrix},$$

$$R_{21} = \begin{bmatrix} D_{12} & L_{21} \\ L_{12} & D_{12} \end{bmatrix}, \quad R_{22} = \begin{bmatrix} D_{11} & L_{11} \\ L_{11} & D_{11} \end{bmatrix}.$$

Remark 9 (about connection with problems of optimal control). Let us consider a boundary value problem with the parameter g for the linear system of SDE

$$dx_g(t) = A(t)x_g(t)dt + g\Sigma(t)dw(t),$$
$$x_g(\alpha) = x_\alpha, \qquad x_g(\beta) = x_\beta.$$

Using the well-known results of the theory of optimal control (see, for example, [61]), it is easy to show that for $g \to 0$ the solutions $x_g(t)$ of this boundary value problem converge to the solution $x(t)$ of the following problem of optimal control:

$$\frac{d}{dt}x(t) = A(t)x(t) + \Sigma(t)u(t),$$
$$x(\alpha) = x_\alpha, \qquad x(\beta) = x_\beta,$$
$$\int_\beta^\alpha \left[\sum_i u_i^2(t)\right] dt \to \min .$$

Task B.5. Formulate and prove rigorously the statement from Remark 9.

Task B.6. Investigate conditions of non-singularity of the matrix $K(t)$, $t > 0$, (cf. Remark 3) in terms of the matrices A and Σ.

Task B.7. Study the boundary value problems for the stochastic heat equation $x''(t) - bx(t) = \sigma dw(t)$. Analyse the correctness of different boundary conditions.

Task B.8. Develop algorithms and software for solving boundary value problems for linear systems of SDE. Make use of: a) the formulas for conditional distributions and Markov property for solutions of SDE; b) the method of Green's functions. Under which conditions the deterministic methods can be applied for solving stochastic boundary value problems?

Task B.9. Construct numerical methods for the solution of boundary value problems for non-linear SDE.

Appendix C

On interpolation of positive definite functions and stationary random sequences

Here we shall obtain representations for stationary random processes and correlation functions, which relate the values on an integer grid with those on a real axis. These representations enable us to interpolate stationary random sequences and positive definite functions of a discrete argument retaining properties of stationarity and positive definiteness, respectively. In addition, we shall consider generalizations for random fields which can be useful in processing geophysical measurements.

1. Let us consider a random process $x(t)$, which is stationary in the wide sense, with mean zero, continuous correlation function

$$R(s) = Mx(t+s)\bar{x}(t),$$

and spectral measure $F(d\lambda)$:

$$R(t) = \int_{-\infty}^{+\infty} \exp(it\lambda)F(d\lambda),$$

$$F(\Delta) = \lim_{T\to\infty} \frac{1}{2\pi} \int_{-T}^{T} \frac{e^{-ibt} - e^{-iat}}{-it} R(t)\,dt$$

for

$$\Delta = (a,b), \quad -\infty \le a < b \le +\infty, \quad F(\{a\}) = F(\{b\}) = 0.$$

By $w(d\lambda)$ we denote stochastic spectral measure of the process $x(t)$:

$$w(\Delta) = \lim_{T\to\infty} \frac{1}{2\pi} \int_{-T}^{T} \frac{e^{-ibt} - e^{iat}}{-it} x(t)\,dt,$$

$$M|dw(\lambda)|^2 = F(d\lambda), \quad x(t) = \int_{-\infty}^{+\infty} \exp(it\lambda)\,dw(\lambda).$$

The random sequence $x_n = x(n)$ is a stationary one with mean zero, correlation function $R(n)$, spectral measure

$$F_1(d\lambda) = \sum_{k=-\infty}^{+\infty} F(d(\lambda + 2k\pi)), \quad d\lambda \subset [-\pi, \pi), \tag{C.1}$$

$$F_1(\Delta) = \frac{1}{2\pi}(b-a)R(0) + \lim_{N\to\infty} \frac{1}{2\pi} \sum_{\substack{n=-N \\ n\neq 0}}^{N} \frac{e^{-ibn} - e^{-ian}}{-in} R(n)$$

and stochastic spectral measure

$$w_1(d\lambda) = \sum_{k=-\infty}^{+\infty} w(d(\lambda + 2k\pi)), \quad \lambda \in [-\pi, \pi), \tag{C.2}$$

$$w_1(\Delta) = \frac{1}{2\pi}(b-a)x_0 + \lim_{N\to\infty} \frac{1}{2\pi} \sum_{\substack{n=-N \\ n\neq 0}}^{N} \frac{e^{-ibn} - e^{-ian}}{-in} x_n,$$

where $-\pi \leq a < b < \pi$, $F_1(\{a\}) = F_1(\{b\}) = 0$. We assume that there exist spectral densities $f(\lambda)$ and $f_1(\lambda)$ ($F(d\lambda) = f(\lambda)d\lambda$, $F_1(d\lambda) = f_1(\lambda)d\lambda$) such that

$$f(\lambda) = (2\pi)^{-1} \int_{-\infty}^{+\infty} \exp(-it\lambda)R(t)\,dt, \quad \lambda \in (-\infty, +\infty), \tag{C.3}$$

$$f_1(\lambda) = (2\pi)^{-1} \sum_{n=-\infty}^{+\infty} \exp(-i\lambda n)R(n), \quad \lambda \in [-\pi, \pi), \tag{C.4}$$

$$f_1(\lambda) = \sum_{k=-\infty}^{+\infty} f(\lambda + 2k\pi), \quad \lambda \in [-\pi, \pi). \tag{C.5}$$

Transformation of the spectrum (C.1), (C.5), while changing a continuous parameter to a discrete one, is called *the aliasing effect.* The formula of aliasing is dual of the Poisson formula (see, e.g. [25]):

$$\sum_{k=-\infty}^{+\infty} R(t + 2k\pi) = \sum_{n=-\infty}^{+\infty} \exp(itn)f(n).$$

2. We denote by $\widetilde{F}(d\lambda)$ the σ-finite measure in $(-\infty, +\infty)$, which is obtained by periodical extension of measure $F_1(d\lambda)$ from the interval $[-\pi, +\pi)$ to the whole real axis. It is obvious that the measure $F(d\lambda)$ is absolutely continuous with respect to the measure $\widetilde{F}_1(d\lambda)$. By the Radon-Nikodym theorem there exists a density $\frac{dF}{d\widetilde{F}_1}(\lambda)$ that is determined uniquely up to equivalence with respect to measure $\widetilde{F}_1(d\lambda)$. We denote

$$g(\lambda) = \frac{1}{2\pi}\frac{dF}{d\widetilde{F}_1}(\lambda).$$

We assume now that the measure $F(d\lambda)$ and, hence, the measure $\widetilde{F}_1(d\lambda)$ are absolutely continuous with respect to Lebesgue measure:

$$F(d\lambda) = f(\lambda)\,d\lambda, \quad F_1(d\lambda) = f_1(\lambda)\,d\lambda, \quad \widetilde{F_1}(d\lambda) = \tilde{f}_1(\lambda)\,d\lambda.$$

Under this assumption we have

$$g(\lambda) = \frac{1}{2\pi}\frac{f(\lambda)}{\tilde{f}_1(\lambda)}. \tag{C.6}$$

for λ such that $\tilde{f}_1(\lambda) \neq 0$. For the points λ where $\tilde{f}_1(\lambda) = 0$ we define the function $g(\lambda)$ so that it remains nonnegative and

$$\sum_{k=-\infty}^{+\infty} g(\lambda + 2k\pi) = \frac{1}{2\pi}. \tag{C.7}$$

Thus, evidently,

$$\int_{-\infty}^{+\infty} g(\lambda)d\lambda = 1.$$

Let us consider a positive definite function

$$K(t) = \int_{-\infty}^{+\infty} \exp(it\lambda)g(\lambda)\,d\lambda \tag{C.8}$$

and assume that

$$g(\lambda) = (2\pi)^{-1}\int_{-\infty}^{+\infty} \exp(-it\lambda)K(t)\,dt. \tag{C.9}$$

Then Fourier transform converts

$$f(\lambda) = 2\pi g(\lambda)\tilde{f}_1(\lambda)$$

to the equality

$$R(t) = \sum_{n=-\infty}^{+\infty} K(t-n)R(n)$$

and, in view of (C.7), we have

$$K(0) = 1, \qquad K(m) = 0 \quad \text{for} \quad m = \pm 1, \pm 2, \dots. \tag{C.10}$$

Indeed,

$$\begin{aligned} f(\lambda) &= 2\pi g(\lambda)\tilde{f}_1(\lambda) \\ &= (2\pi)^{-1} \sum_{n=-\infty}^{+\infty} R(n)\exp(-i\lambda n) \int_{-\infty}^{-\infty} \exp(-i\lambda t)K(t)\,dt \\ &= (2\pi)^{-1} \sum_{n=-\infty}^{+\infty} R(n) \int_{+\infty}^{-\infty} \exp(-i\lambda t)K(t-n)\,dt \\ &= (2\pi)^{-1} \int_{+\infty}^{-\infty} \exp(-i\lambda t)\Big(\sum_{n=-\infty}^{+\infty} K(t-n)R(n)\Big)\,dt. \end{aligned} \tag{C.11}$$

The above computations are correct if equalities (C.3), (C.4), (C.9) hold.

Definition. The spectral density $f(\lambda)$ is said to be *regular* density if equalities (C.3), (C.4) hold for almost all λ, and there exists function $g(\lambda)$ defined by (C.6), (C.7), such that (C.9) is fulfilled almost surely.

Different regularity conditions for spectral density can be obtained on the basis of the theory of Fourier series and transforms (see, e.g. [20], [118]). For most applications it is sufficient to note that a piecewise differentiable spectral density $f(\lambda)$ is regular if the function $\sum_{k=-\infty}^{+\infty} f(\lambda + 2k\pi)$ is also piecewise differentiable in $(-\pi, \pi)$.

Thus, the following assertion becomes obvious.

Statement C.1. *If $R(t)$ is correlation function of a stationary process $x(t)$ with regular spectral density, then the following representation is fulfilled*

$$R(t) = \sum_{n=-\infty}^{+\infty} K(t-n)R(n). \tag{C.12}$$

where $K(t)$ is a continuous positive definite function satisfying condition (C.10). Moreover, if the spectral density of the stationary sequence $x(n)$ does not equal to zero almost everywhere, then this representation is unique.

Corollary. *An arbitrary positive definite sequence $R(n)$ can be extended to correlation function $R(t)$ of a stationary process with regular spectral density by equation (C.12), where $K(t)$ is a continuous positive definite function that satisfies the conditions (C.7)–(C.10).*

Let us consider the following example. If

$$R(t) = \begin{cases} 1 - |t|/T, & |t| \leq T; \\ 0, & |t| > T; \end{cases}$$

with integer T, then

$$\begin{aligned}
f(\lambda) &= \frac{T}{2\pi}\left(\frac{\sin(T\lambda/2)}{T\lambda/2}\right)^2 = \frac{1-\cos(T\lambda)}{\pi T \lambda^2}, \\
\tilde{f}_1(\lambda) &= (2\pi T)^{-1}\left(\frac{\sin(T\lambda/2)}{\sin(\lambda/2)}\right)^2, \\
K(t) &= \begin{cases} 1 - |t|, & |t| < 1; \\ 0, & |t| \geq 1; \end{cases} \\
g(\lambda) &= (2\pi)^{-1}\left(\frac{\sin(\lambda/2)}{\lambda/2}\right)^2 = \pi^{-1}(1-\cos\lambda)\lambda^{-2}, \\
f(\lambda) &= 2\pi g(\lambda)\tilde{f}_1(\lambda).
\end{aligned}$$

The definition of regularity introduced above places a number of constraints on spectral density. Note that we can give up condition (C.3). Indeed, it is known [113] that in general case we have convergence

$$f(\lambda) = \lim_{\varepsilon\to 0}(2\pi)^{-1}\int_{-\infty}^{+\infty} \exp(-it\lambda)R(t)e^{-\varepsilon|t|}\,dt, \quad \lambda \in (-\infty, +\infty),$$

in L^1-norm and almost surely. However, the other regularity conditions (C.4), (C.6), (C.7) ensure (C.11) and hence (C.3).

In addition, we can abandon the constraint (C.4). For spectral density $f_1(\lambda)$ we have representation with Cesaro sum (see, e.g. [20])

$$f_1(\lambda) = \lim_{N\to\infty} (2\pi)^{-1} \sum_{n=-N}^{N} \exp(-i\lambda n)\Big(1 - \frac{|n|}{N+1}\Big) R(n),$$

$$\lambda \in (-\pi, +\pi).$$

However, computations (C.11) remain valid with Cesaro summing as well. Thus, the following assertion holds.

Statement C.2. A. *Let $R(t)$ be correlation function of a stationary process with spectral density $f(\lambda)$. We assume that there exists a nonnegative function $f(\lambda)$ that is defined by relations (C.6), (C.7) and satisfies (C.9). Then we have*

$$R(t) = \sum_{n=-\infty}^{-\infty} K(t-n)R(n)$$

with summing in Cesaro sense.

B. *Let $R(n)$ be correlation function of a stationary sequence with spectral density $f_1(\lambda)$. We consider a nonnegative function $g(\lambda)$ such that conditions (C.7)–(C.9) are fulfilled. Then the series*

$$B(t) = \sum_{n=-\infty}^{+\infty} K(t-n)R(n) \tag{C.13}$$

is summable in Cesaro sense for all $t \in (-\infty, +\infty)$. Moreover, $K(t)$ satisfies (C.10), function $B(t)$ is continuous positive definite and $B(n) = R(n)$. Spectral density $f(\lambda)$ corresponding to correlation function $B(t)$ is determined by

$$f(\lambda) = 2\pi g(\lambda)\tilde{f}_1(\lambda), \quad \lambda \in (-\infty, +\infty), \tag{C.14}$$

where $\tilde{f}_1(\lambda)$ is the periodic extension of the spectral density $f_1(\lambda)$ from interval $[-\pi, \pi)$ to the whole real axis.

Remark. A possibility to use representations (C.13), (C.14) for extending positive definite sequences to positive definite functions of a continuous argument was considered in [83].

3. In the previous item we obtained the expressions that relate values of positive definite functions on the integer grid with those on the whole real axis. Attempts to obtain analogous relations for stationary random processes lead to the following result.

Statement C.3. *Let N denote an open set, in which spectral measure of a stationary random process $x(t)$ is concentrated. If the condition*

$$\{N + 2k\pi\} \cap N = \emptyset \quad \textit{for all} \quad k = \pm 1, \pm 2, \ldots \tag{C.15}$$

is fulfilled, then we have representation for the process $x(t)$

$$x(t) = \sum_{n=-\infty}^{+\infty} A(t-n)x(n), \tag{C.16}$$

where the series converges in mean square and $A(t)$ is a continuous positive definite function for which

$$A(0) = 1, \qquad A(n) = 0 \quad \textit{for} \quad n = \pm 1, \pm 2, \ldots. \tag{C.17}$$

Representation (C.16) is unique if the sequence $x(n)$ has spectral density that is nonzero almost everywhere. In this case,

$$A(t) = (2\pi)^{-1} \int_{-\infty}^{+\infty} \exp(it\lambda)a(\lambda)\, d\lambda, \tag{C.18}$$

where $a(\lambda)$ is the indicator of support of the spectral measure of the random process $x(t)$.

Proof. We denote by L_x^2 a closure of a linear span of random variables $x(t)$, $t \in \mathbb{R}$, and by L_F^2 a Hilbert space of complex-valued functions with norm

$$\|\Phi\|^2 = \int_{\mathbf{R}} |\Phi(\lambda)|^2 F(d\lambda),$$

where $F(d\lambda)$ is a spectral measure of the process $x(t)$. Due to isometry of spaces L_x^2 and L_F^2 (see, for example, [31]) representation (C.16) follows from the equality

$$e^{i\lambda t} = \sum_{n=-\infty}^{+\infty} A(t-n)e^{i\lambda n}, \quad \lambda \in N. \tag{C.19}$$

Using (C.15) and (C.18), we obtain (C.19) as follows

$$\begin{aligned} e^{i\lambda t} &= \int_N e^{i\lambda' t}\delta(\lambda - \lambda')\,d\lambda' \\ &= \int_N e^{i\lambda' t}(2\pi)^{-1} \sum_{n=-\infty}^{\infty} e^{i(\lambda-\lambda')n}\,d\lambda' \\ &= \sum_{n=-\infty}^{\infty} e^{i\lambda n}(2\pi)^{-1} \int_N \exp(i\lambda'(t-n))\,d\lambda' \\ &= \sum_{n=-\infty}^{\infty} e^{i\lambda n}(2\pi)^{-1} \int_{-\infty}^{+\infty} \exp(i\lambda'(t-n))a(\lambda')\,d\lambda' \\ &= \sum_{-\infty}^{+\infty} A(t-n)e^{i\lambda n}, \quad \lambda \in N, \end{aligned}$$

where $\delta(x) = \frac{1}{2\pi}\sum_n \exp(inx)$ is the periodic δ-function and $a(\lambda)$ is the indicator of the set N.

Property (C.17) is equivalent to the fact that equality

$$\sum_{k=-\infty}^{+\infty} a(\lambda + 2k\pi) = 1, \tag{C.20}$$

holds almost everywhere, while positive definiteness of $A(t)$ means that the function $a(\lambda)$ is nonnegative. We can meet these requirements by extending the set N so that condition (C.15) is not violated.

If the stationary sequence $x(n)$ has spectral density that is nonzero almost everywhere, then the set N, which contains the support of the spectrum and satisfies conditions (C.15) and (C.20), is unique. In this case the representation (C.16) is equivalent to equality (C.19). Since the correspondence between the functions $a(\lambda)$ and $A(t)$ is one-to-one, the representation (C.16) is unique.

The statement has been proved.

Remark. The Kotelnikov–Shannon formula for a stationary process with spectrum concentrated in the interval $(-\pi, \pi)$ is a special case of representation (C.16):

$$x(t) = \sum_{n=-\infty}^{+\infty} \frac{\sin(\pi(t-n))}{\pi(t-n)} x(n).$$

Example. If spectrum of a stationary process $x(t)$ is concentrated in the set $(-2\pi, -\pi) \cup (\pi, 2\pi)$, then representation (C.16) takes place with

$$A(t) = \frac{\sin(2\pi t) - \sin(\pi t)}{\pi t}.$$

Statement C.4. *Assume that for a stationary (in the wide sense) random process $x(t)$ with absolutely continuous spectral measure the following representation takes place*

$$x(t) = \sum_{n=-\infty}^{+\infty} A(t-n) x(n), \tag{C.21}$$

$$A(t) = (2\pi)^{-1} \int_{-\infty}^{+\infty} \exp(it\lambda) a(\lambda)\, d\lambda,$$

where $a(\lambda)$ is a bounded, piecewise continuous and absolutely integrable function on $(-\infty, +\infty)$. Then there exists an open set N satisfying condition (C.15) in which spectral measure of the process $x(t)$ is concentrated. If, in addition, spectral density of the sequence x_n is nonzero almost everywhere, then condition (C.17) is fulfilled and representation (C.21) is unique.

Proof. Let N be an open set which contains spectrum of the process $x(t)$. Representation (C.21) implies that for all $\lambda \in N$ and t we have

$$\begin{aligned} e^{i\lambda t} &= \sum_{n=-\infty}^{+\infty} A(t-n) e^{i\lambda n} = \\ &= \sum_{n=-\infty}^{+\infty} e^{i\lambda n} (2\pi)^{-1} \int_{-\infty}^{+\infty} \exp(i(t-n)\gamma) a(\gamma)\, d\gamma = \\ &= \int_{-\infty}^{+\infty} e^{it\gamma} (2\pi)^{-1} \sum_{n=-\infty}^{\infty} e^{in(\lambda-\gamma)} a(\gamma)\, d\gamma = \\ &= e^{it\lambda} \sum_{k=-\infty}^{+\infty} a(\lambda + 2k\pi), \end{aligned}$$

and, hence,

$$\sum_{k=-\infty}^{+\infty} a(\lambda + 2k\pi) = 1, \quad \forall \lambda \in N.$$

It remains to use the following lemma.

Lemma. *Suppose that a stationary process $y(t)$ and a stationary sequence x_n with absolutely continuous spectral measure are connected by the relation*

$$y(t) = \sum_{n=-\infty}^{\infty} A(t-n)x_n,$$

$$A(t) = (2\pi)^{-1} \int_{-\infty}^{\infty} \exp(it\lambda)a(\lambda)\, d\lambda,$$

where $a(\lambda)$ is a bounded, piecewise continuous and absolutely integrable function on $(-\infty, +\infty)$.

Then stochastic spectral measure $dw(\lambda)$ of the process $y(t)$ is determined in terms of stochastic spectral measure $dw_1(\lambda)$ of the sequence x_n by equation

$$dw(\lambda + 2k\pi) = a(\lambda + 2k\pi)dw_1(\lambda), \quad \lambda \in (-\pi, +\pi),$$

Moreover, for almost all $\lambda \in \mathbb{R}$ the following condition is fulilled: $a(\lambda) \neq 0$ imlies $a(\lambda + 2k\pi) = 0$ for $k = 1, 2, \ldots$.

Proof of the Lemma follows from computations

$$w[c,d] = \frac{1}{2\pi} \int_{-\infty}^{\infty} \frac{e^{-idt} - e^{-ict}}{-it} \sum_n A(t-n)x_n dt =$$

$$= \frac{1}{2\pi} \int_{-\infty}^{\infty} \sum_n \int_{-\infty}^{\infty} \frac{1}{2\pi} \frac{e^{-idt} - e^{-ict}}{-it} \exp(i(t-n)\lambda)a(\lambda)x_n d\lambda dt =$$

$$= \frac{1}{2\pi} \sum_n \int_{-\infty}^{\infty} \exp(-in\lambda) I_{[c,d]}(\lambda)a(\lambda)x_n d\lambda =$$

$$= \frac{1}{2\pi} \sum_n \int_{-\infty}^{\infty} \int_{-\pi}^{\pi} \exp(i(\gamma - \lambda)n) I_{[c,d]}(\lambda) a(\lambda) dw_1(\gamma) d\lambda =$$

$$= \frac{1}{2\pi} \sum_k \int_{-\pi}^{\pi} \int_{-\pi}^{\pi} \sum_n \exp(i(\gamma - \lambda)n) I_{[c,d]}(\lambda + 2k\pi) a(\lambda + 2k\pi) dw_1(\gamma) d\lambda =$$

$$= \sum_k \int_{-\pi}^{\pi} I_{[c,d]}(\lambda + 2k\pi) a(\lambda + 2k\pi) dw_1(\lambda).$$

Here we denote by $I_{[c,d]}(\lambda)$ the indicator of the set $[c, d]$. In the latter equality we use the fact that series $\frac{1}{2\pi} \sum_n \exp(inx)$ defines the periodic δ-function.

4. The results presented above can be readily generalized to the case of multidimensional parameter. In particular, in order to interpolate correlation functions of random fields defined on the integer grid in $\mathbb{R}^d$, we can use the following assertion.

Statement C.5. *We denote by R_n a correlation function on the integer grid in $\mathbb{R}^d$ (index $n \in \mathbb{R}^d$ denotes here all possible d-dimensional vectors with integer components), and by $f(\lambda)$ the corresponding spectral density:*

$$R_n = \int_{[-\pi,\pi]^d} \exp(i < n, \lambda >) f(\lambda) d\lambda,$$

$$f(\lambda) = (2\pi)^{-d} \sum_n \exp(-i < n, \lambda >) R_n, \quad \lambda \in [-\pi, +\pi)^d,$$

where $< .,. >$ denotes the scalar product in $\mathbb{R}^d$.

Let $K(t)$ be a correlation function with spectral density $g(\lambda)$,

$$K(t) = \int_{\mathbb{R}^d} \exp(i < t, \lambda >) g(\lambda) d\lambda, \quad t \in \mathbb{R}^d,$$

$$g(\lambda) = (2\pi)^{-d} \int_{\mathbb{R}^d} \exp(-i < t, \lambda >) K(t) dt, \quad \lambda \in \mathbb{R}^d,$$

such that

$$K(0) = 1, \qquad K(n) = 0 \quad \text{for} \quad n \neq 0, \tag{C.22}$$

and the series $\sum_n K(t-n)R_n$ *converges uniformly. By* $\tilde{f}(\lambda)$ *we denote periodic extension of the spectral density* $f(\lambda)$ *from* $[-\pi, \pi)^d$ *to the space* $\mathbb{R}^d$. *Then the function*

$$B(t) = \sum_n K(t-n)R_n, \quad t \in \mathbb{R}^d,$$

is a correlation (positive definite) function with spectral density

$$h(\lambda) = (2\pi)^d g(\lambda)\tilde{f}(\lambda),$$

Moreover,

$$B(n) = R_n$$

and property (C.22) is equivalent to equality

$$\sum_n g(\lambda + 2n\pi) = (2\pi)^{-d}, \quad \lambda \in [-\pi, \pi]^d.$$

Analogues of Statements C.1–4 and Lemma from Item 3 for the case of multidimensional parameter can be formulated and proved in the obvious way. In multidimensional case condition (C.15) has the form

$$\{N + 2k\pi\} \cap N = \emptyset$$

for an arbitrary nonzero d-dimensional vector k with integer components and representation (C.16) can be written as

$$x(t) = \sum_n A(t-n)x(n), \quad t \in \mathbb{R}^d,$$

where summation is made over all d-dimensional vectors with integer components.

Appendix D

Coding of multiplicative pseudorandom number generators

D.1. Multiplicative generators: description and coding

The "multiplicative generator" is one of the most popular modelling algorithms of a sequence of pseudorandom numbers α_j uniformly distributed on the interval $(0,1)$:

$$\alpha_j = C * n_j, \quad n_{j+1} = n_j * L \ (\mathrm{mod}\, M), \quad C = M^{-1} \tag{D.1}$$

The method was proposed by P.H. Lehmer in 1951.

Appendix is devoted to specific features of realization of multiplicative generators by high-level programming languages. Though a low-level computer code can be much faster, generators realized by high-level languages are sometimes more preferable because of their universality and portability.

The following procedures written in Turbo Pascal for IBM PC are presented below:

RAND_1: $M = 2^{31}$, $L = 331\,804\,469$.

RAND_2: $M = 2^{40}$, $L = 5^{17}$ [4], [22].

RAND_3: $M = 2^{55}$, $L = 48\,044\,648\,058\,128\,333$ [4].

RAND_4: $M = 2^{70}$, $L = 5^{29}$.

The procedure RAND_5 realizes modulo-1 addition of two multipicative generators [6]:

$$\alpha_j = C_1 * n_j^1 + C_2 * n_j^2 \ (\mathrm{mod}\, 1),$$
$$n_{j+1}^1 = n_j^1 * L_1 (\mathrm{mod}\, M_1), \quad C_1 = M_1^{-1},$$
$$n_{j+1}^2 = n_j^2 * L_2 (\mathrm{mod}\, M_2), \quad C_2 = M_2^{-1},$$
$$L_1 = 1\,528, \ M_1 = 1\,400\,159, \ L_2 = 1\,356, \ M_2 = 1\,500\,419.$$

The procedures RAND_6, RAND_7, RAND_8 also realize modulo-1 addition of several multipicative generators. We shall call such generators as **complex multiplicative generators**. The complex multiplicative generators, realized in RAND_5 – RAND_8, were proposed by M.V. Antipov.

The original information on pseudo-random number generators may be found in [22], [34], [46], [50], [105], [111]. The results of verification of some multiplicative and mixed congruential generators are presented in [4]. In this paper, there was found a set of multipliers with the best functional-correlation characteristics. In monograph [74], it is recommended that the multiplicative generator with coefficients

$$M = 2^m, \quad L = 5^{2p+1}, \quad p = \max\{q : 5^{2q+1} < M\} \tag{D.2}$$

be used. This consideration was taken into account for choosing the multiplier for the procedure RAND_4.

To attain the maximal period of a generator, the modulus and the multiplier should be chosen in a special way. The following statements can be found in [21], [105].

Theorem.

1. *The maximal period of multiplicative generator (D.1) with modulus $M = 2^m$, $m \geq 3$, is equal to 2^{m-2}. The maximal period is attained for all odd n_0 if $L \bmod 8 = 3$ or 5.*

2. *A multiplicative generator has period $M - 1$ only if M is prime. Then the period divides $M - 1$, and is $M - 1$ if and only if L is a primitive root , that is, $L \neq 0$ and $L^{(M-1)/p} \neq 1 \bmod M$ for each prime factor p of $M - 1$.*

Restrictions on the quantity of pseudo-random numbers, simulated in a numerical algorithm, were studied in [5]. In particular, it was recommended there that no more than $P^{2/3}$ pseudo-random numbers be used, where P is a period of generator.

A large period is necessary but not sufficient for high statistical quality of a generator. A well-known example of a poor generator with sufficiently large period 2^{29} is RANDU (this generator was widely used on IBM 360/370 and PDP-11 machines):

$$M = 2^{31}, \quad L = 65539 = 2^{16} + 3. \tag{D.3}$$

Below we briefly describe the procedures RAND_1-RAND_8.

Procedure RAND_1 makes use of specific features of the overflow treatment for integer numbers in Turbo Pascal (the range of integer numbers of LONGINT type is $-2^{31} \ldots 2^{31}-1$).

Procedures RAND_2-RAND_4 virtually implement multiplication of integer numbers in 2^{14} base notation:

$$
\begin{array}{ccccccccc}
 & & & * & a_n & \cdots & a_3 & a_2 & a_1 \\
 & & & & b_n & \cdots & b_3 & b_2 & b_1 \\
\hline
 & & + & c_{1,n+1} & c_{1,n} & \cdots & c_{1,3} & c_{1,2} & c_{1,1} \\
 & + & c_{2,n+1} & c_{2,n} & c_{2,n-1} & \cdots & c_{2,2} & c_{2,1} & \\
 & & \cdots & \cdots & \cdots & \cdots & \cdots & \cdots & \cdots \\
+ & c_{n,n+1} & \cdots & \cdots & c_{n,1} & & & & \\
\hline
 & a'_{2n} & \cdots & \cdots & a'_n & \cdots & a'_3 & a'_2 & a'_1,
\end{array}
$$

where a_i, b_i and c_{ij} are digits of the corresponding number system.

To make the algorithm faster, specific operations of Turbo Pascal are used in procedures RAND_2-RAND_4:

$$C \bmod 2^k = C \text{ and } (2^k - 1), \quad C \text{ div } 2^k = C \text{ shr } k.$$

Here **and** is the operation of bitwise multiplication and **shr** is the bitwise shift-right operation.

Note that the algorithm realized in procedures RAND_2–RAND_4 may be easily tuned to any modulus $M = 2^m$ and an arbitrary multiplier L. Constants $b_1, \ldots, b_n$ in these procedures present decomposition of the multiplier L in 2^k $(k \le 14)$ base notation:

$$L = b_1 + b_2 2^k + b_3 2^{2k} + \ldots + b_n 2^{(n-1)k},$$

where n is the number of digits of decomposition. The 2^{14} base number system may be used in the algorithm for $m \le 98$, $n \le 7$ (for $n > 7$, overflows over the range of LONGINT type would occur; for $k = 13$, the values $n \le 31$ are admissible).

In the procedures RAND_5-RAND_8, pseudo-random numbers are simulated by modulo-1 summation of several multiplicative generators (see [6], [7]). The moduli of multiplicative generators are different Smith's numbers, while the multipliers are primitive roots for the corresponding

Smith's numbers (a Smith's number is a prime that can be represented in the form $2P+1$, where P is prime, too). The period of such a multiplicative generator is equal to the corresponding Smith's number minus one, and the period of the complex multiplicative generator is equal to the product of the periods of the multiplicative generators.

Exercise D.1. Find the values of $5^p \bmod 8$ for different p.

Exercise D.2. Prove that for the generator `RANDU` (D.3), the pseudorandom sequence satisfies the following property

$$n_{k+2} = 6n_{k+1} - 9n_k \pmod{2^{31}}.$$

Exercise D.3. Program the multiplicative generators with the following parameters:

a) $M = 1400423, \quad L = 1499$ [4] (show that in this case, M is the Smith's number, and L is a primitive root for M);

b) $M = 2^{31}, \quad L = 5^{13}$ [112];

c) $M = 2^{42}, \quad L = 5^{17}$ [34], [111];

d) $M = 2^{55}, \quad L = 14\,312\,160\,467\,264\,461$ [4];

e) $M = 2^{54}, 2^{55}, 2^{56}, \quad L = 5^{23}$;

f) $M = 2^{98}$, choose L according to (D.2);

g) $M = 2^{48}, \quad L = 44\,485\,709\,377\,909$ [105];

h) $M = 2^{59}, \quad L = 5^{13}$ [105].

Find periods of these generators.

Exercise D.4. Study the "real-valued" implementation of the multiplicative algorithm realized in the procedure `RAND_MCU` (see the next Item and [27]). Write a similar procedure for the generator with modulus $M = 2^{40}$ and multiplier $L = 5^{17}$. Compare the speed of the "real-valued" implementation of the generator with the speed of the "integer-valued" procedure `RAND_2`.

D.2. Procedures in Pascal, Fortran and C

```
{
 Subject: Demonstration program for different
          generators of pseudo-random numbers
 Author: Prigarin Sergei Mikhailovich
   e-mail: smp@osmf.sscc.ru
   http://osmf.sscc.ru/~smp
 Institution: Institute of Computational Mathematics
              and Mathematical Geophysics (Novosibirsk),
              Siberian Branch of Russian Academy
              of Sciences
 Language: Turbo Pascal for IBM PC
}

{$A+,B-,D-,E+,F-,G+,I+,L-,N+,O-,P-,Q-,R+,S+,T-,V-,X+}
{$M 16384,0,655360}
PROGRAM RND_TEST;

{
  Multiplicative generator of pseudo-random numbers
          M = 2**31,  L = 331 804 469
         (period  2**29 = 536 870 912)

  Seed:  1 <= N <= 2 147 483 647
}
function RAND_1:real;
const N:longint=1;
      L=331804469;
      C=1.0/MaxLongInt;   {MAXLONGINT = 2 147 483 647}
begin
  N := N * L;
  if N <= 0  then  N := N + MaxLongInt + 1;
  RAND_1 := N * C;
end;

{
  Multiplicative generator of pseudo-random numbers
       M = 2**40,  L = 5**17 = 762 939 453 125
         (period 2**38 = 274 877 906 944)
```

```
 Initial values:  1<=a1<=16383, 0<=a2<=16383,
                  0<=a3<=4095,  a1 is odd !
}
function  RAND_2: real;
const
  a1:longint=1; a2:longint=0; a3:longint=0;
  b1:longint=11973; b2:longint=2800; b3:longint=2842;
  x1=1.0/4096.0; x2=x1/16384.0; x3=x2/16384.0;
var c11, c12, c13, c21, c22, c31: longint;
    d1, d2, d3: longint;
begin
  c11 := b1*a1;  c12 := b1*a2;  c13 := b1*a3;
  c21 := b2*a1;  c22 := b2*a2;
  c31 := b3*a1;
  d1 := c11;
  d2 := c21 + c12 + (d1 shr 14);
  d3 := c31 + c22 + c13 + (d2 shr 14);
  a1:=d1 and 16383; a2:=d2 and 16383; a3:=d3 and 4095;
{ for standard Pascal:
  d2 := c21 + c12 + (d1 div 16384);
  d3 := c31 + c22 + c13 + (d2 div 16384);
  a1:=d1 mod 16384; a2:=d2 mod 16384; a3:=d3 mod 4096;
}
  RAND_2 := a3*x1 + a2*x2 + a1*x3;
END;

{
  Multiplicative generator of pseudo-random numbers
     M = 2**55,  L = 48 044 648 058 128 333
     (period 2**53 = 9 007 199 254 740 992)

 Initial values:      1<=a1<=16383,  0<=a2<=16383,
      0<=a3<=16383,  0<=a4<=8191,   a1 is odd !
}
function RAND_3: real;
const
  a1:longint=1;  a2:longint=0;  a3:longint=0;
                              a4:longint=0;
  b1: longint=1997;  b2: longint=4980;
```

```
  b3: longint=1445;  b4: longint=10924;
  x1 = 1.0/8192.0;  x2 = x1/16384.0;
  x3 = x2/16384.0;  x4 = x3/16384.0;
var c11,c12,c13,c14,c21,c22,c23,c31,c32,c41: longint;
    d1,d2,d3,d4: longint;
begin
  c11:=b1*a1; c12:=b1*a2; c13:=b1*a3; c14:=b1*a4;
  c21:=b2*a1; c22:=b2*a2; c23:=b2*a3;
  c31:=b3*a1; c32:=b3*a2;
  c41:=b4*a1;
  d1 := c11;
  d2 := c21 + c12 + (d1 shr 14);
  d3 := c31 + c22 + c13 + (d2 shr 14);
  d4 := c41 + c32 + c23 + c14 + (d3 shr 14);
  a1:=d1 and 16383;  a2:=d2 and 16383;
  a3:=d3 and 16383;  a4:=d4 and 8191;
  RAND_3 := a4*x1 + a3*x2 + a2*x3 + a1*x4;
end;

{
  Multiplicative generator of pseudo-random numbers
 M = 2**70, L = 5**29 = 186 264 514 923 095 703 125
   (period 2**68 = 295 147 905 179 352 825 856)

 Initial values:  0 <= a1,a2,a3,a4,a5 <= 16383,
                           a1 is odd !
}
function RAND_4: real;
const
  a1:longint=1;  a2:longint=0;  a3:longint=0;
                 a4:longint=0;  a5:longint=0;
  b1:longint=15957; b2:longint=11842; b3:longint=5942;
                    b4:longint=15391; b5:longint=2584;
  x1=1.0/16384.0;  x2=x1/16384.0;  x3=x2/16384.0;
                   x4=x3/16384.0;  x5=x4/16384.0;
var c11,c12,c13,c14,c15,c21,c22,c23,c24,
    c31,c32,c33,c41,c42,c51: longint;
    d1,d2,d3,d4,d5: longint;
begin
  c11:=b1*a1; c12:=b1*a2; c13:=b1*a3; c14:=b1*a4;
```

```
  c15:=b1*a5;
  c21:=b2*a1; c22:=b2*a2; c23:=b2*a3; c24:=b2*a4;
  c31:=b3*a1; c32:=b3*a2; c33:=b3*a3;
  c41:=b4*a1; c42:=b4*a2;
  c51:=b5*a1;
  d1 := c11;
  d2 := c21 + c12 + (d1 shr 14);
  d3 := c31 + c22 + c13 + (d2 shr 14);
  d4 := c41 + c32 + c23 + c14 + (d3 shr 14);
  d5 := c51 + c42 + c33 + c24 + c15 + (d4 shr 14);
  a1:=d1 and 16383;  a2:=d2 and 16383;
  a3:=d3 and 16383;  a4:=d4 and 16383;
  a5:=d5 and 16383;
  RAND_4 := a5*x1 + a4*x2 + a3*x3 + a2*x4 + a1*x5;
end;

{
 Pseudo-random number generator:  modulo-1
 addition of several multiplicative generators
 (period 1400158*1500418=2 100 822 266 044)

   Initial values:  1 <= n1 <=1400158,
                    1 <= n2 <=1500418
}
function RAND_5: real;
const  n1:longint=1;  n2:longint=1;
const  c1=1.0/1400159;  c2=1.0/1500419;
var  R: real;
begin
  n1 := (n1*1528) mod 1400159;
  n2 := (n2*1356) mod 1500419;
  R := n1 * c1 + n2 * c2;
  if (R > 1.0)  then  RAND_5 := R - 1.0  else
  RAND_5 := R;
end;


{
  Pseudo-random number generator:  modulo-1
  addition of several multiplicative generators
```

```
 (period 1400158*1500418*1001002=2.10292729e+18)

   Initial values:  1 <= n1 <= 1400158,
                    1 <= n2 <= 1500418,
                    1 <= n3 <= 1001002
}
function RAND_6: real;
const n1:longint=1; n2:longint=1; n3:longint=1;
const c1=1.0/1400159; c2=1.0/1500419; c3=1.0/1001003;
var  R: real;
begin
  n1 := (n1*1528) mod 1400159;
  n2 := (n2*1356) mod 1500419;
  n3 := (n3*1856) mod 1001003;
  R := n1 * c1 + n2 * c2 + n3 * c3;
  RAND_6 := frac(R);
end;

{
      Pseudo-random number generator:  modulo-1
    addition of several multiplicative generators
 (period 27000706*24001798*21000982=1.3610011717e+22)

   Initial values:  1 <= n1 <= 27000706,
                    1 <= n2 <= 24001798,
                    1 <= n3 <= 21000982
}
FUNCTION Rand_7: double;
Function DMOD(x,y:double):double;
Begin
  DMOD:=x-y*int(x/y);
End;
const n1:double=1; n2:double=1; n3:double=1;
const c1:double=1.0/27000707;
      c2:double=1.0/24001799;
      c3:double=1.0/21000983;
var  R: double;
begin
  n1 := DMOD(n1*6173, 27000707);
  n2 := DMOD(n2*7853, 24001799);
```

```
  n3 := DMOD(n3*10330, 21000983);
  R := n1 * c1 + n2 * c2 + n3 * c3;
  if (R > 2.0)  then  RAND_7 := R - 2.0  else
  if (R > 1.0)  then  RAND_7 := R - 1.0  else
  RAND_7 := R;
end;

{
 Pseudo-random number generator:  modulo-1
 addition of several multiplicative generators
 (period 27000706*24001798*21000982*
       *15001606*10000222=2.0417656605e+36)

   Initial values:  1 <= n1 <= 27000706,
                    1 <= n2 <= 24001798,
                    1 <= n3 <= 21000982,
                    1 <= n4 <= 15001606,
                    1 <= n5 <= 10000222
}
FUNCTION Rand_8: double;
Function DMOD(x,y:double):double;
Begin
  DMOD:=x-y*int(x/y);
End;
const n1:double=1; n2:double=1; n3:double=1;
      n4:double=1; n5:double=1;
const c1:double=1.0/27000707;
      c2:double=1.0/24001799;
      c3:double=1.0/21000983;
      c4:double=1.0/15001607;
      c5:double=1.0/10000223;
var  R: double;
begin
  n1 := DMOD(n1*6173, 27000707);
  n2 := DMOD(n2*7853, 24001799);
  n3 := DMOD(n3*10330, 21000983);
  n4 := DMOD(n4*20203, 15001607);
  n5 := DMOD(n5*45773, 10000223);
  R := n1*c1 + n2*c2 + n3*c3 + n4*c4 + n5*c5;
  RAND_8 := frac(R);
```

```
end;

{
  Multiplicative generator of pseudo-random numbers
      uniformly distributed between 0 and 1:

  This random number generator is used in the package
  "MCU-RFFI" (see http://www.mcu.com.ru)

    n(0) = 2001
    n(i+1) = mod(n(i)*(5**15), 2**47)
    rand(i+1) = n(i+1) / (2**47)
}
FUNCTION Rand_MCU:double;
Var
  U,V:double;
Const
  D24:double=16777216.0;                         {2^24}
  D47:double=140737488355328.0;                  {2^47}
  A1:double=1818.0;
  A2:double=16599437.0;                {5^15=D24*A1+A2}
  X1:double=0.0;  X2:double=2001.0;

Function DMOD(x,y:double):double;
Begin
  DMOD:=x-y*int(x/y);
End;

Begin
  U:=X2*A2;   V:=X1*A2+X2*A1;
  V:=DMOD(V,D24);   V:=D24*V+U;
  V:=DMOD(V,D47);
  X1:=Int(V/D24);   X2:=V-D24*X1;
  Rand_MCU:=V/D47;
End;

{}
var  test, code: integer;  i,k: longint;  x: double;
BEGIN  {PROGRAM}
  k:=0;
```

```
  val(paramstr(1), test, code);
  case test of
    1:  for i:=1 to 1000000 do  x := Rand_1;
    2:  for i:=1 to 1000000 do  x := Rand_2;
    3:  for i:=1 to 1000000 do  x := Rand_3;
    4:  for i:=1 to 1000000 do  x := Rand_4;
    5:  for i:=1 to 1000000 do  x := Rand_5;
    6:  for i:=1 to 1000000 do  x := Rand_6;
    7:  for i:=1 to 1000000 do  x := Rand_7;
    8:  for i:=1 to 1000000 do  x := Rand_8;
    9:  for i:=1 to 1000000 do  x := Rand_MCU;
    else for i:=1 to 1000000 do  x := Random;
{ the built-in generator for Turbo Pascal 6.0:
         Random = abs(n(j) / 2^31),
  n(j) = [134775813*n(j-1)+1+2^31] mod 2^32 - 2^31
}
  end;
  writeln(x);
END.
```

```
C -----------------------------------------------------
C   Multiplicative generator of pseudo-random numbers
C        uniformly distributed between 0 and 1:
C  modulus  2**40,  multiplier  5**17 = 762 939 453 125
C       (cycle length  2**38 = 274 877 906 944)
C
C  The generator was extensively used (and was
C  carefully tested) in the Department of Stochastic
C  Simulation in Physics of the Novosibirsk Computing
C  Center; see, for example, the book
C
C    "Stochastic  Simulation",
C    by S.M.Ermakov and G.A.Mikhailov,
C    Moscow, Nauka, 1982
C
C  Fortran realization, by S.M.Prigarin (smp@osmf.sscc.ru)
C
C  (INTEGER values are assumed to be 32-bit integers !)
```

```
C  (the result RND may be of single or double precision)
C
C  Initial values:  0 <= A1 <= 16383,  0 <= A2 <= 16383,
C                   0 <= A3 <= 4095,   A1 must be odd !
C  Here we set:  A1=1, A2=0, A3=0
C ----------------------------------------------------------
      SUBROUTINE  RANDOM(RND)
      INTEGER  a1, a2, a3
      INTEGER  b1, b2, b3
      INTEGER  c11, c12, c13, c21, c22, c31
      INTEGER  d1, d2, d3
C
      DOUBLE PRECISION  RND
C
      DOUBLE PRECISION  x1, x2, x3
      SAVE  a1, a2, a3
      SAVE  b1, b2, b3
      SAVE  x1, x2, x3
      DATA  a1/1/, a2/0/, a3/0/
      DATA  b1/11973/, b2/2800/, b3/2842/
      DATA  x1/0.000244140625D0/
      DATA  x2/1.4901161194D-8/
      DATA  x3/9.0949470178D-13/

      c11 = b1 * a1
      c12 = b1 * a2
      c13 = b1 * a3
      c21 = b2 * a1
      c22 = b2 * a2
      c31 = b3 * a1
      d1 = c11
      d2 = c21 + c12 + d1/16384
      d3 = c31 + c22 + c13 + d2/16384
      a1 = mod(d1,16384)
      a2 = mod(d2,16384)
      a3 = mod(d3,4096)

      RND = a3*x1 + a2*x2 + a1*x3
      return
      END
```

```
/*
 Subject: Demonstration program for generator with 'jumps'
 Author: Prigarin Sergei Mikhailovich
   e-mail: smp@osmf.sscc.ru
   http://osmf.sscc.ru/~smp
 Institute: Institute of Computational Mathematics
            and Mathematical Geophysics (Novosibirsk),
            Siberian Branch of Russian Academy of Sciences
 Language: Turbo C for IBM PC
*/

#include <stdio.h>
#include <math.h>

double Random1(void);

double Rand1_Frog(long ind, long s1, long s2, long K);

/* Main function  */
int main()
{
  long i;  double x,y;

  printf("Testing of generator with 'jumps': \n");

  x=Rand1_Frog(0,0,1,1000);
  printf(" 0 %23.16e\n",x);

  for( i=1; i<=1000; ++i )
  {  x = Rand1_Frog(1,0,0,0);  }
  printf(" 1000 %23.16e",x);

  x=Rand1_Frog(2,0,0,0);
  printf(" %23.16e\n",x);

  for( i=1; i<=2000; ++i )
  {  x = Rand1_Frog(1,0,0,0);  }
  printf(" 3000 %23.16e",x);
```

```
  x=Rand1_Frog(2,0,0,0);
  x=Rand1_Frog(2,0,0,0);
  printf(" %23.16e",x);

  x=Rand1_Frog(0,0,1,3000);
  x=Rand1_Frog(2,0,0,0);
  printf(" %23.16e\n",x);

  x=Rand1_Frog(0,0,1,100000);
  x=Rand1_Frog(2,0,0,0);
  printf(" 100000 %21.16e",x);

  for( i=1; i<=100000; ++i )
  {  x = Rand1_Frog(1,0,0,0);  }
  printf(" %23.16e",x);

  for( i=1; i<=100000; ++i )
  {  x=Random1();  }
  printf(" %23.16e\n",x);

  return(0);
}

/*

  Multiplicative generator of pseudo-random numbers
      uniformly distributed between 0 and 1:
 module  2**40,  multiplier  5**17 = 762 939 453 125
     (cycle length  2**38 = 274 877 906 944)

       n(0) = 1     /or =a1+a2*2^14+a3*2^28/
     n(i+1) = mod(n(i)*(5^17), 2^40)
   rnd(i+1) = n(i+1) / (2^40)

  "C" realization, by S.M.Prigarin (smp@osmf.sscc.ru)

 Initial values:  1 <= a1 <= 16383,  0 <= a2 <= 16383,
          0 <= a3 <= 4095,   a1 must be odd !
 Here we set:  a1=1, a2=0, a3=0
```

```
*/
double Random1(void)
{
  static long a1=1, a2=0, a3=0;          /* initial values */
  const static long b1=11973, b2=2800, b3=2842;
  long c11, c12, c13, c21, c22, c31, d1, d2, d3;

  const double x1=1.0/4096.0, x2=x1/16384.0, x3=x2/16384.0;

  c11 = b1*a1;  c12 = b1*a2;  c13 = b1*a3;
  c21 = b2*a1;  c22 = b2*a2;
  c31 = b3*a1;
  d1 = c11;
  d2 = c21 + c12 + (d1 >> 14);
  d3 = c31 + c22 + c13 + (d2 >> 14);
    a1=d1 & 16383;  a2=d2 & 16383;  a3=d3 & 4095;
/*  a1=d1 % 16384;  a2=d2 % 16384;  a3=d3 % 4096;*/
  return(a3*x1 + a2*x2 + a1*x3);
}

/*
-----------------------------------------------------------
          Multiplicative generator with "jumps"
     modulus  2**40,  multiplier  5**17 = 762 939 453 125
              (period 2**38 = 274 877 906 944)

  N(i+1) = mod(N(i)*(5**17), 2**40)
  a(i+1) = N(i+1) / (2**40)
-----------------------------------------------------------
 The function simulates sequences of pseudo-random
 numbers

 a(0), a(1), a(2), a(3), ...

 uniformly distributed in the interval (0,1).

 In addition, it can simulate sequences with "jumps"
 of the length K

 a(0), a(K), a(2K), a(3K), ...
```

```
 This technique is often used in stochastic
 simulation to realize the method of dependent tests.
 ==========================================================
 Description of the function:

 double Rand1_Frog(long ind, long s1, long s2, long K)

 0) Rand1_Frog(0,s1,s2,K) - initialization of the generator:
 s1,s2 define the value Rand1_Frog=a(i)=a(j*K) for i=0, j=0,
 and preparation for "jumps" of the length K
 (Turbo C for IBM PC:   K <= MaxLongInt = 2 147 483 647).

 Initial values: 0 <= s1 <= 2^20-1,
                 1 <= s2 <= 2^20-1 (must be odd!).

 If you do not use initialization of Rand1_Frog,
 then s1=0, s2=1, K=1 by default.

 1) Rand1_Frog(1,s1,s2,K) - simulation of the next
 pseudo-random number:  i=i+1, Rand1_Frog=a(i)
 /the values of s1,s2,K are ignored/.

 2) Rand1_Frog(2,s1,s2,K) - simulation of the next
 "jump":  j=j+1, i=j*K, Rand1_Frog=a(i)=a(j*K)
 /the values of s1,s2,K are ignored/.
 ====================================================
 S.M.Prigarin (smp@osmf.sscc.ru)
*/
double Rand1_Frog(long ind, long s1,long s2, long K)
{
  double U,V;  long i;

  const double D20=1048576.0;          /* 2^20 */
  const double D40=1099511627776.0;    /* 2^40 */
  const double A1=727595.0;
  const double A2=798405.0;    /* 5^17=D20*A1+A2 */
  static double Aj1=727595.0;
  static double Aj2=798405.0;  /* 5^17=D20*Aj1+Aj2 */
  static double X1=0.0;    static double X2=1.0;
```

```
  static double Xj1=0.0;   static double Xj2=1.0;

  if (ind==1)
  {
    U=X2*A2;    V=X1*A2+X2*A1;
    V=fmod(V,D20);    V=D20*V+U;    V=fmod(V,D40);
    X1=floor(V/D20);    X2=V-D20*X1;
    return(V/D40);
  }
  else
  if (ind==2)
  {
    U=Xj2*Aj2;    V=Xj1*Aj2+Xj2*Aj1;
    V=fmod(V,D20);    V=D20*V+U;    V=fmod(V,D40);
    Xj1=floor(V/D20);   Xj2=V-D20*Xj1;
    return(V/D40);
  }
  else;
  {
    X1=s1; X2=s2;  Xj1=s1; Xj2=s2;

    Aj1=A1;  Aj2=A2;
    for (i=2; i<=K; ++i)
    {
      U=Aj2*A2;    V=Aj1*A2+Aj2*A1;
      V=fmod(V,D20);    V=D20*V+U;    V=fmod(V,D40);
      Aj1=floor(V/D20);   Aj2=V-D20*Aj1;
    };

    V=X1*D20+X2;
    return(V/D40);
  };
}
```

Bibliography

[1] A. Albert. Regression and the Moor-Penrose Pseudoinverse. Academic Press, New York, 1972

[2] T. Anderson. An Introduction to Multivariate Statistical Analysis. John Wiley & Sons Inc., New York, 1958

[3] T. Anderson. The Statistical Analysis of Time Series. John Wiley & Sons Inc., New York, 1971

[4] M.V. Antipov. Assessment and selection of pseudorandom number generator for a given class of problems. Preprint No. 554, Comp. Cent. Sib. Division, Russian Acad. Sci., Novosibirsk, 1985 [in Russian]

[5] M.V. Antipov. Restrictions on numerical Monte Carlo methods caused by using pseudo-random numbers. In: Theory and Applications of Statistical Modelling, Comp. Cent. Sib. Division, USSR Acad. Sci., Novosibirsk, 1989, P. 3–9 [in Russian]

[6] M.V. Antipov. Collections of pseudorandom number generators for personal computers. Preprint No. 910, Comp. Cent. Sib. Division, Russian Acad. Sci., Novosibirsk, 1990 [in Russian]

[7] M.V. Antipov and G.A. Mikhailov. On the improvement in random number generators by using a modulo 1 sum. Russian Journal of Numerical Analysis and Math. Modelling (1996) 11, No. 2, P. 93–111

[8] S.R. Anvarov and S.M. Prigarin. Numerical simulation of the spatio-temporal structure of the sea swell surface in optical problems. Optika Atm. Okeana (1994) 7, No. 5, P. 685-693 [in Russian]

[9] H. Bateman and A. Erdelyi. Higher Transcendental Functions. McGraw-Hill Book Company, Inc., New York, 1953

[10] H. Bateman and A. Erdelyi. Tables of Integral Transfoms, Vol.2. McGraw-Hill Book Company, Inc., New York, 1954

[11] R. Benktus et al. Test examples of time series. In: Applications of Probability Theory and Mathematical Statistics, Vilnius, 1985, Issue 6, P. 31-46 [in Russian]

[12] O.V. Besov, V.P. Il'in, and S.M. Nikolski. Integral Representations and Embedding Theorems. Nauka, Moscow, 1996 [in Russian]

[13] P. Billingsley. Convergence of Probability Measures. John Wiley & Sons, Inc., New York, 1968

[14] A.A. Borovkov. Convergence of measures and random processes. Uspehi Mat. Nauk (1976) 31, No. 2, P. 3-68 [in Russian]

[15] E.P. Box and G.M. Jenkins. The Time Series Analysis. Holdenday, San Francisco, 1970

[16] N.A. Buglanova. Convergence of approximate models of random fields. Ph. D. Thesis, Almaty, 1995 [in Russian]

[17] Yu.V. Bulavskii, Minimization of the variance of the collision estimate in the problems with alternating sign elements. Zh. Vychisl. Mat. Mat. Fiz (1984) 24, No.1, P. 148-152 [in Russian]

[18] I.M. Davidan, L.I. Lopatukhin and V.A. Rozhkov. Wind-Driven Undulation in the World Ocean. Gidrometeoizdat, Leningrad, 1985 [in Russian]

[19] L. Devroye. Non-uniform random variate generation. N.Y.: Springer, 1986.

[20] R.E. Edwards. Fourier Series. A Morden Introduction. Springer-Verlag, New-York, 1979.

[21] S.M. Ermakov. Monte Carlo methods and Related Questions. Nauka, Moscow, 1975 [in Russian]

[22] S.M. Ermakov and G.A. Mikhailov. Statistical Modelling. Nauka, Moscow, 1982 [in Russian]

[23] S.M. Ermakov and A.A. Zhiglyavskii. Mathematical Theory of the Optimal Experiment. Nauka, Moscow, 1987 [in Russian]

[24] J. Feder. Fractals. Plenum Press, New York, 1988.

[25] W. Feller. An Introduction to Probability Theory and its Applications, Vol.2. John Wiley & Sons, Inc., New York, 1971

[26] Applications of Fractional Calculus in Physics (Ed. R. Hilfer). World Scientific, Singapore, 1999

[27] A.D. Frank-Kamenetskiy. Simulation of Neutron Trajectories for Computing Reactors by Monte Carlo Method. Atomizdat, Moscow, 1978 [in Russian]

[28] A.S. Frolov and N.N. Chentsov. The use of dependent sampling in Monte Carlo method to obtain smooth curves. In: Proceedings of the VI All-Union Conference on the Theory of Probability and Mathematical Statistics, Gos. Izd. Polit. i Nauchn. Lit. LitSSSR, Vilnyus, 1962, P. 425-437 [in Russian]

[29] L.S. Gandin and R.L. Kagan. Statistical Methods of Interpretation of Meteorological Data. Gidrometeoizdat, Leningrad, 1967 [in Russian]

[30] I.I. Gikhman and A.V. Skorokhod. Theory of Random Processes, Vol.1. Nauka, Moscow, 1971 [in Russian]

[31] I.I. Gikhman and A.V. Skorokhod. An Introduction to the Theory of Random Processes. Nauka, Moscow, 1977 [in Russian]

[32] S.K. Godunov. Ordinary Differential Equations with Constant Coefficients. Boundary Value Problems. Novosibirsk State Univ. Press, 1994 [in Russian]

[33] S.K. Godunov. Modern Aspects of Linear Algebra. Scientific Book, Novosibirsk, 1997 [in Russian]

[34] D.I. Golenko. Simulation and Statistical Analysis of Pseudorandom Numbers by Electronic Computers. Nauka, Moscow, 1965 [in Russian]

[35] M.L. Grigorjeva and E.I. Ostrovskii. Computation of integrals of discontinuous functions by the method of dependent tests. Zh. Vychisl. Mat. Mat. Fiz. (1996) 36, No.12, P. 28-38 [in Russian]

[36] Z.A. Grikh. Statistical modelling of homogeneous isotropic random fields and isotropic random fields on a sphere. Ph. D. Thesis, Kiev, 1987 [in Russian]

[37] Handbook on Probability Theory and Mathematical Statistics (Eds. V.S. Korolyuk, N.I. Portenko, A.V. Skorokhod, A.F. Turbin). Nauka, Moscow, 1985 [in Russian]

[38] E.J. Hannan. Multiple Time Series. John Wiley & Sons, Inc., New York, 1970

[39] N.A.J. Hastings and J.B. Peacock. Statistical Distributions. Butterworth, London, 1975

[40] S. Heinrich. A multilevel version of the method of dependent tests. In: Mathematical methods in stochastic simulation and experimental design. Proceedings of the 3d St.Petersburg Workshop on simulation (1998, June 28 - July 3). Publishing House of Saint Petersburg University, 1998, P. 31-35

[41] S. Heinrich. The multilevel method of dependent tests. Interner Bericht / Universitaet Kaiserslautern; 300/99, 1999

[42] S. Heinrich S. and E. Sindambiwe. Monte Carlo complexity of parametric integration. Journal of complexity (1999) 15, P. 317-341

[43] H.D. Ikramov. Numerical Solution of Matrix Equations. Nauka, Moscow, 1984 [in Russian]

[44] A.V. Ivanov. On convergence of distributions of functionals of measurable random fields. Ukrainskiy Matem. Zh. (1980) 32, No. 1, P. 27-34 [in Russian]

[45] N.C. Jain and M.B. Marcus. Central limit theorems for $C(S)$-valued random variables. J.Funct.Anal. (1975) 19, No. 3, P. 216-231

[46] B. Jansson. Random Number Generators. Almquist and Wiksell, Stockholm, 1966

[47] R.R. Kanter and S.M. Prigarin. Numerical modelling of the wind driven undulation for investigation of the field of reflected optical radiation. Preprint No. 829. Compt. Cent. Sib. Division, USSR Acad. Sci., Novosibirsk, 1989 [in Russian]

[48] B.A. Kargin and S.M. Prigarin. Modelling of stochastic fields of heap clouds and study of their radiative properties with Monte Carlo method. Preprint No. 817. Compt. Cent. Sib. Division, USSR Acad. Sci., Novosibirsk, 1988 [in Russian]

[49] B.A. Kargin and S.M. Prigarin. Simulation of sea undulation surface and study of its optical properties by Monte Carlo method. Optika Atm. Okeana (1992) 5, No. 3, P. 285-291 [in Russian]

[50] D.E. Knuth. The Art of Computer Programming. Vol.2. Seminumerical Algorithms. Reading, Massachusetts, 1969

[51] A.N. Kolmogorov. Theory of Information and Theory of Algorithms. Nauka, Moscow, 1987 [in Russian]

[52] Yu.V. Kozachenko and M.I. Yadrenko. Local properties of sample functions of random fields. In: Theory of Probability and Mathematical Statistics, Vol.14, Kiev, 1976, P. 53-66 [in Russian]

[53] R.H. Kraichnan. Diffusion by a random velocity field. Phys. Fluids (1970) 13, No. 1, P. 22-31

[54] V.M. Kruglov. Additional Chapters of Probability Theory. Vysshaya Shkola, Moscow, 1984 [in Russian]

[55] Yu.M. Krylov. Spectral Methods of Investigation and Calculation of Wind-Driven Waves. Gidrometeoizdat, Leningrad, 1966 [in Russian]

[56] O.A. Kurbanmuradov. Functional convergence of Monte Carlo approximations of homogeneous Gaussian random fields. Preprint No. 999, Comp. Cent. Sib. Division, Russian Acad. Sci., Novosibirsk, 1993 [in Russian]

[57] O.A. Kurbanmuradov. Convergence of numerical models for the Gaussian fields. Russian Journal of Numerical Analysis and Math. Modelling (1995) 10, No. 4, P. 311-324

[58] O.A. Kurbanmuradov, K.K. Sabelfeld and G. Chopanov. Statistical modelling of the impurity diffusion in random velocity fields. Random field modelling. Preprint No. 775, Comp. Cent. Sib. Division, USSR Acad. Sci., Novosibirsk, 1988 [in Russian]

[59] H.J. Kushner. An application of the Sobolev imbedding theorems to criteria for the continuity of process with a vector parameter. Ann. Math. Statist. (1969) 40, No. 2, P. 517-526

[60] H.J. Kushner. Probability Methods for Approximation in Stochastic Control. Academic Press, New York, 1977

[61] Yu.K. Lando. Elements of Mathematical Theory of Optimal Movement Control. Prosveshenie, Moscow, 1984.

[62] P. Lankaster. Theory of Matrices. Academic Press, New York, 1969

[63] V.A. Lebedev. Martingales, Convergence of Probability Measures and Stochastic Equations. MAI Press, Moscow, 1996 [in Ruassian]

[64] N.N. Leonenko and A.V. Ivanov. Statistical Analysis of Random Fields. Visha Shkola, Kiev, 1986 [in Russian]

[65] I.A. Malyshav and Yu.I. Palagin. Mathematical modelling of multidimensional random fields. Tekhnicheskaia Kibernetika (1980), No. 6, P. 10–20 [in Russian]

[66] A.S. Marchenko and V.A. Ogorodnikov. Simulation of Gaussian sequences of large length with an arbitrary correlation function. Zh. Vychisl. Mat. i Mat. Fiz. (1984) 24, P. 1514–1519 [in Russian]

[67] A.S. Marchenko and V.A. Ogorodnikov. Autoregressive processes with a given correlation structure. Izv. Vuzov. Matematika (1989) 7, P. 63–67 [in Russian]

[68] G.I. Marchuk, G.A. Mikhailov, M.A. Nazaraliev, R.A. Darbinian, B.A. Kargin, B.S. Elepov. Monte Carlo Methods in Atmosperic Optics. Springer-Verlag, Berlin, Heidelberg, New York, 1989

[69] S.L. Marple, Jr. Digital Spectral Analysis with Applications. Mir, Moscow, 1990 [in Russian]

[70] G.A. Mikhailov. Some Problems of the Theory of Monte Carlo Methods. Nauka, Novosibirsk, 1974

[71] G.A. Mikhailov. Numerical construction of a random field with given spectral density. Dokl. Akad. Nauk SSSR (1978) 238, No. 4, P. 793–795 [in Russian]

[72] G.A. Mikhailov. Approximate models of random processes and fields. Zh. Vychisl. Mat. Mat. Fiz. (1983) 23, No. 3, P. 558–566 [in Russian]

[73] G.A. Mikhailov. Minimization of Computational Costs of Non-Analogue Monte Carlo Methods. World scientific, Singapore, 1991

[74] G.A. Mikhailov. Optimization of Weighted Monte Carlo Methods. Springer-Verlag, Berlin, 1992

[75] G.A. Mikhailov. New Monte Carlo Methods with Estimating Derivatives. VSP, Utrecht, 1995

[76] G.A. Mikhailov and K.K. Sabelfeld. On numerical modelling of the impurity diffusion in stochastic velocity fields. Izv. Akad. Nauk SSSR, Fiz. Atm. Okean (1980) 16, No. 3, P. 229–235 [in Russian]

[77] O.A. Mikheev. Study of criteria of optimality of stochastic estimators in the method of dependent tests for calculating integrals depending on a parameter. Graduation Thesis. Novosibirsk State University, Novosibirsk, 1993 [in Russian]

[78] Yu.-A.R. Mullamaa. Covering of the sky by cumulus clouds. In: Radiatsya i Oblachnost, IFA Akad. Nauk ESSR, Tartu, 1969 [in Russian]

[79] Yu.-A.R. Mullamaa, M.A. Sulev, V.K. Pyldmaa, et al. Stochastic Structure of Clouds and Radiation Fields. IFA Akad. Nauk ESSR, Tartu, 1972 [in Russian]

[80] A.M. Obukhov. Statistically homogeneous fields on a sphere. Uspehi Mat. Nauk (1947) 2, No. 2, P. 196–198 [in Russian]

[81] V.A. Ogorodnikov. On dynamic-stochastic forecasting. Izv. Akad. Nauk SSSR, FAO (1975) 11, No. 8, P. 851–853 [in Russian]

[82] Ogorodnikov V.A. Statistical simulation of discrete random processes and fields. Sov. J. Num. Anal. and Math. Modelling (1990) 5, No. 6, P. 489–509

[83] V.A.Ogorodnikov and S.M.Prigarin. Numerical Modelling of Random Processes and Fields: Algorithms and Applications. VSP, Utrecht, 1996, 240p.

[84] Yu.I. Palagin. Mathematical modelling of vector-valued random fields. Avtomatika i Telemekhanika (1981), No. 2, P. 35–42 [in Russian]

[85] Yu.I. Palagin, S.V. Fedotov and A.S. Shalygin. Probabilistics modelling of the fields of atmospheric turbulence and sea undulation in studying complex systems. Radiotekhnika i Elektronika (1986), No. 4, P. 721–729 [in Russian]

[86] Yu.I. Palagin, S.V. Fedotov and A.S. Shalygin. Parametric models for statistical modelling of vector inhomogeneous random fields. Avtomatika i Telemekhanika (1990), No. 6, P. 79–89 [in Russian]

[87] V.I. Paulauskas and A.Yu. Rachkauskas. Accuracy of Approximation in Central Limit Theorem for Banach Spaces. Mokslas, Vilnyus, 1987 [in Russian]

[88] S.M. Prigarin. On weak convergence of approximate models of Gaussian random fields. In: Theory and Applications of Statistical Modelling, Comp. Cent. Sib. Division, USSR Acad. Sci., Novosibirsk, 1988, P. 31–39 [in Russian]

[89] S.M. Prigarin. Some applications of Jane-Marcus theorem to statistical modelling. In: Methods of Statistical Modelling, Comp. Cent. Sib. Division, USSR Acad. Sci., Novosibirsk, 1990, P. 103–108 [in Russian]

[90] S.M. Prigarin. Asymptotic efficiency of estimators in the method of dependent tests. In: Theses of VIII All-Union Conference "Monte Corlo Methods in Computational Mathematics and Mathematical Physics". Novosibirsk, 1991, P. 94–96 [in Russian]

[91] S.M. Prigarin. Spectral models of vector homogeneous fields. Preprint No. 945. Compt. Cent. Sib. Division, USSR Acad. Sci., Novosibirsk, 1992 [in Russian]

[92] S.M. Prigarin. Convergence of numerical models of random fields in Monte Carlo methods. Russian Journal of Numerical Analysis and Math. Modelling (1992) 7, No. 5, P. 441–456

[93] S.M. Prigarin. Weak convergence of probability measures in spaces of continuously differentiable functions. Sibir. Mat. Zh. (1993) 34, No. 1, P. 140–144 [in Russian]

[94] S.M. Prigarin. Spectral models of vector-valued random fields. Bull. NCC. Ser. Num. Anal. Issue 4. (1993), P. 45–60

[95] S.M. Prigarin. Conditional spectral models of Gaussian homogeneous fields. Russian Journal of Numerical Analysis and Math. Modelling (1998) 13, No. 1, P. 57–68

[96] S.M. Prigarin. Essay on numerical modelling of random fields. Doctor Thesis (second dissertation). Novosibirsk, 1998, 313p. [in Russian]

[97] S.M. Prigarin. On convergence and optimization of functional Monte Carlo estimators in Sobolev's spaces. Sibirski zhurnal vychislitelnoi matematiki (1999) 2, No. 1, P. 57–67 [in Russian]

[98] S.M. Prigarin. Introduction to Numerical Modelling of Random Processes and Fields. Novosibirsk State Univ. Press, 1999, 301p. [in Russian]

[99] S.M. Prigarin, B.A. Kargin, and U.G. Oppel. Random fields of broken clouds and their associated direct solar radiation, scattered transmission, and albedo. Pure and Applied Optics, A., (1998) 7, No.6, P. 1389–1402

[100] Yu.V. Prokhorov. Convergence of random processes and limit theorems of probability theory. Teoriya Veroyatnostey i ee Primeneniya (1956) 1, No. 2, P. 178–238 [in Russian]

[101] Yu.V. Prokhorov and Yu.A. Rozanov. Probability Theory. Basic Notions, Limit Theorems, Random Processes. Nauka, Moscow, 1973 [in Russian]

[102] Radiation in Cloudy Atmospere (Ed. E.M. Feigelson). Gidrometeoizdat, Leningrad, 1981 [in Russian]

[103] C.R. Rao. Linear statistical inference and its applications. John Wiley & Sons. Inc., New York [Russian transl., Nauka, Moscow, 1968]

[104] F. Riesz and B.Sz.-Nady. Lectures on Functional Analysis. Inostrannaya Literatura, Moscow, 1979 [in Russian]

[105] B.D. Ripley. Stochastic simulation. Wiley, New York, 1987

[106] K.K. Sabelfeld. Monte Carlo methods in boundary value problems. Springer-Verlag, Berlin, 1991.

[107] K.K. Sabelfeld and O.A. Kurbanmuradov. Numerical statistical model of classical incompressible isotropic turbulence. Sov. J. Numer. Anal. Math. Modelling (1990) 5, No. 3, P. 251–263

[108] A.A. Samarskiy. Introduction into Numerical Methods. Nauka, Moscow, 1987 [in Russian]

[109] A.F. Sizova and T.M. Tovstik. Simulation of a random field on a sphere. Vestnik LGU, 1984, No. 1, P. 118–120 [in Russian]

[110] A.S. Shalygin and Yu.I. Palagin. Applied Methods of Statistical Modelling. Mashinostroenie, Leningrad, 1986 [in Russian]

[111] I.M. Sobol. Numerical Monte Carlo Methods. Nauka, Moscow, 1973 [in Russian]

[112] I.M. Sobol. Monte Carlo Method. Nauka, Moscow, 1985 [in Russian]

[113] E.M. Stein and G. Weiss. Introduction to Fourier Analysis on Euclidean Spaces. Princeton University Press, Princeton, New Jersey, 1971.

[114] A.A. Sveshnikov. Applied Methods of the Theory of Random Functions. Nauka, Moscow, 1968 [in Russian]

[115] T.M. Tovstik. Simulation of homogeneous Gaussian field. Proceedings of Symposium "Methods of Representation and Apparat. Anal. of Random Processes and Fields", Leningrad, 1978, P. 75–77 [in Russian]

[116] V.I. Tikhonov and M.A. Mironov. Markov Processes. Sov. Radio, Moscow, 1977 [in Russian]

[117] A.N. Tikhonov and A.A. Samarskiy. Equations of Mathematical Physics. Nauka, Moscow, 1977 [in Russian]

[118] E.C. Titchmarsh. Introduction to the Theory of Fourier Integrals. Clarendon Press, Oxford, 1962.

[119] V.V. Voevodin and Yu.A. Kuznetsov. Matices an Computations. Nauka, Moscow, 1984 [in Russian]

[120] A.V. Voitishek. Randomized numerical spectral model of stationary random function. In: Mathematical Simulation Models of Systems, Comp. Cent. Sib. Division, USSR Acad. Sci., Novosibirsk, 1983, P. 17–25 [in Russian]

[121] A.V. Voytishek and S.M. Prigarin. On functional convergence of estimates and models in Monte Carlo methods. Zh. Vychisl. Mat. Mat. Fiz. (1992) 32, No.10, P. 1641–1651 [in Russian]

[122] R.J. Wilson. Weak convergence of probability measures in spaces of smooth functions. Stochastic Processes and Their Applications (1986) 23, P. 333–337

[123] M.I. Yadrenko. Isotropic random fields of Markov type in Euclidean and Hilbert spaces. In: Proceedings of the All-Union Conference on the Theory of Probability and Mathematical Statistics, Yerevan, 1960, P. 425–437 [in Russian]

[124] M.I. Yadrenko. Spectral Theory of Random Fields. Kiev Univ. Press, Kiev, 1980 [in Russian]

[125] A.M. Yaglom. Some classes of random fields in n-dimensional space, related to stationary random processes. Teoriya Veroyatnostey i ee Primeneniya (1957) 2, No. 3, P. 292–337 [in Russian]

[126] V.V. Yurinskiy. On sample continuity of random fields. Teoriya Veroyatnostey i ee Primeneniya (1973) 18, No. 3, P. 633–639 [in Russian]

[127] V.M. Zolotarev. Modern Theory of Summation of Independent Random Variables. Nauka, Moscow, 1986 [in Russian]

Notation

$A \Delta B = (A - B) \cup (B - A)$	symmetric difference of sets A and B;
A^+	pseudoinverse of a matrix A;
A^T	transpose of a matrix A;
$a \vee b$	maximum of numbers a, b;
$\mathbb{C}$	the set of complex numbers;
$\mathbf{D}\xi$	variance of a random variable ξ;
$\mathcal{D}_{m_1 \ldots m_k}, \mathcal{D}^m$	operators of differentiation;
$\mathrm{Dev}(\xi, F)$	F-deviation of an estimator ξ;
$\mathrm{diam}(A) = \sup\limits_{x,y \in A} \lvert x - y \rvert$	diameter of a set A;
$\dim(X)$	dimension of a space X;
I_ν, K_ν	the modified Bessel functions (cylindrical functions of imaginary variable), K_ν is the Macdonald function;
$\mathrm{Im}(A)$	image of an operator (a matrix) A;
J_ν	the Bessel function of the first kind of the order ν;
$\mathrm{Ker}(A)$	kernel of an operator (a matrix) A;
$Lo(A)$	linear span of a set A;
$\mathbf{M}\xi$	mathematical expectation of a random variable ξ;
$\mathbf{P}(A), \mathbf{P}\{A\}$	probability of an event A;
$\mathbb{R}$	the set of real numbers;
$\mathrm{Rad}(A) = \lim \lVert A^n \rVert^{1/n}$	spectral radius of an operator A;
$\mathrm{rang}(A)$	rank of a matrix A;
$\mathrm{Re}\, z$, $\mathrm{Im}\, z$	real and imaginary parts of a complex number z;
$\mathrm{sh}\, x = \dfrac{e^x - e^{-x}}{2}$	the hyperbolic sine function;
$\operatorname{vrai\ sup}\limits_{x \in X} f(x)$	the essential upper bound of a function $f(x)$ (this is the smallest number C such that $f(x) \le C$ almost everywhere with respect to a measure on X, by default it is normally assumed to be the Lebesgue measure);
$< a, b >$	the scalar product of vectors a and b.

Index

Bei Fragen zur Produktsicherheit wenden Sie sich bitte an:
If you have any questions regarding product safety,
please contact:

Walter de Gruyter GmbH
Genthiner Straße 13
10785 Berlin
productsafety@degruyterbrill.com